HOME REPAIR AND IMPROVEMENT

MASONRY

HOME REPAIR AND IMPROVEMENT

MASONRY

BY THE EDITORS OF TIME-LIFE BOOKS, ALEXANDRIA, VIRGINIA

The Consultants

Richard Day spent eight years with the Portland Cement Association as a writer and editor on subjects relating to concrete. Based in southern California, he has written numerous articles for *Popular Science* and has authored several books on concrete and masonry.

Dennis Golden, certified by the National Ready Mixed Concrete Association, has been a quality control technician for several concrete companies in the United States. He was also involved in a family concrete contracting business in New Mexico for five years before joining the Public Works Department in Buckley, Washington.

Roger Hopkins has been working in landscape masonry, water features, and granite fabrication since he started Naturalistic Gardens in Sudbury, MA, in the early 1970s. Mr. Hopkins has appeared on PBS programs including *This Old House, The Victory Garden, Nova,* and *The Secrets of Lost Empires*.

Brian E. Trimble is a senior engineer with the Brick Institute of America in Reston, Virginia. Mr. Trimble has written many articles and papers for *The American Ceramic Society Bulletin, ASTM, Progressive Architecture Magazine,* and *The Construction Specifier*. He has also presented numerous seminars on paving, specifications, and computer tools to a variety of groups.

Rhett Whitlock cofounded WDP and Associates Inc. in Manassas, Virginia, in 1995; the company is concerned with masonry design and testing. He received his Doctorate degree in Structural Engineering from Clemson University in 1983. Dr. Whitlock is active in several national masonry organizations including the American Society for Testing and Materials and The Construction Specifications Institute.

CONTENTS

Masonry Techniques

The projects in this book involve techniques that take only a bit of practice to master. Most of them can be learned while you are making repairs to existing brickwork or concrete. In this chapter, we'll show you how to make simple repairs to walls and paving, as well as how to work with masonry.

Replacing bricks →

A Masonry Tool Kit

Working with brick, stone, concrete, and tile requires a number of specialized tools. Pictured below and opposite are the most common ones; others are shown throughout the book. The majority of these tools is available at hardware stores, and some large power tools—a vibratory tamper, for instance—can be rented.

For most masonry projects, you'll also need standard layout and measuring tools such as a 50-foot-long tape measure. A mason's level or long carpenter's level is essential for checking for level and plumb. For laying out corners, you'll need a carpenter's square.

Pointing trowel.
This small trowel is ideal for replacing damaged mortar in a brick wall or shaping mortar joints.

Quarry-tile trowel.
The flat edge is used to spread a bed of mortar for setting tile on a concrete slab. The toothed edge is used to comb the bed; the size of the teeth determines the depth of the bed.

Joint filler.
The long, thin blade is handy for packing mortar into joints in walls and paving.

Mason's trowel.
This large trowel is the basic tool for laying a mortar bed for a brick or block wall as well as for spreading mortar on the units.

Magnesium float.
For smoothing, this lightweight float avoids tearing the surface of air-entrained concrete. For ordinary concrete, a traditional wood float can be used instead.

Edger.
The curved blade rounds the edges of a concrete slab to keep them from chipping.

Jointer.
This tool is used to cut control joints in large concrete slabs, helping to confine cracks along these joints.

Steel trowel.
Used after a float, this trowel imparts an extra-smooth surface to concrete.

Darby.
The long blade is ideal for the initial smoothing of a small concrete slab; some models have two handles.

Brick set.
Strike this tool with a ball-peen hammer to score and cut bricks and stone.

Hawk.
During small repairs to brick, this tool is handy for holding mortar close to the job.

Stone chisel.
Strike this chisel with a ball-peen hammer to score and cut flagstone.

Joint chisel.
This tool is struck with a ball-peen hammer to chip damaged mortar from the joints between bricks. A cold chisel can be used instead.

Ball-peen hammer.
This hammer is for striking metal tools such as brick sets and joint chisels where a standard hammer might be damaged.

Mortar hoe.
This is the traditional tool for mixing mortar and concrete by hand. A shovel can be used instead.

Stonemason's hammer.
The pointed end of this heavy hammer is ideal for trimming small pieces of stone.

Mason's line blocks.
The wood blocks grip the corners of a wall, while the cord forms a guideline for laying brick courses.

Working with Mortar

Mortar is the material that bonds bricks, stones, and blocks together. Once you master mixing and troweling mortar, you can fix walls or paving, or launch more ambitious projects such as building a brick wall or laying a walk.

Preparing the Mix: Mortar consists of Portland cement, sand, hydrated lime, and water. Cement and lime are packaged dry in bags, and the sand should be a clean, fine type called masonry sand—never use beach sand. You can substitute masonry cement (a premeasured mix of cement and lime) for the total amount of cement and lime. Or, buy prepackaged mortar mix; while it is more expensive than buying the separate ingredients, it is more convenient for small jobs. You can also buy colored Portland cement to make mortar that accents joints between bricks or disguises repairs.

Recipes for making 1 cubic foot of dry mortar mix are listed opposite. Type N is suited for most purposes; Type M is best for below-grade work and paving; and Type S is recommended for regions with seismic activity. Mortar can be prepared by hand *(below)*, but you can rent a power mixer for large jobs. Start by preparing 1 cubic foot at a time. You can increase this amount as you learn how much you can comfortably use before the mix hardens.

Spreading Techniques: A mason's trowel is used to lay a bed of mortar for bricks, blocks, or stones, and to spread mortar on these units. While working, stir the mortar often. If it dries out, add water from time to time to restore its workability; but discard mortar that starts to set before it is spread.

Finishing Joints: After the mortar has set but before it hardens—usually under an hour or when it will just hold a thumbprint—you can finish the joints *(pages 14-15)*. Only the joint styles that shed water well are recommended for most outdoor use. Nonwaterproof joints should be reserved for indoor projects or dry climates.

TOOLS

Wheelbarrow
Mortar hoe
Mortar board
Mason's trowel
Mason's level
Convex jointer
V-shaped jointer
Raking tool
Pointing trowel

MATERIALS

Mortar ingredients (Portland cement, masonry sand, hydrated lime)
Bricks or concrete blocks

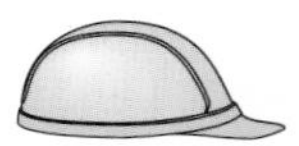

SAFETY TIPS

Mortar—both wet and dry—is caustic. Wear gloves and long sleeves when applying it and add goggles and a dust mask when mixing mortar. Put on hard-toed shoes when handling bricks.

Making a batch of mortar.

At least 4 to 5 gallons of water are generally required for 1 cubic foot of mortar. Be prepared to make adjustments as you test the consistency of the mix.

◆ Measure the dry ingredients into a wheelbarrow, and mix thoroughly with a mortar hoe.

◆ Push the mixture to one end of the wheelbarrow and pour $2\frac{1}{2}$ to 3 gallons of water into the other end. Hoe the dry ingredients into the water. Working back and forth, gradually add more water and mix until the mortar has the consistency of soft mush and there are no lumps.

◆ Test the mix by plowing a curved furrow across the surface with the hoe *(right)*. If the sides of the furrow stay in place and the clinging mortar can be shaken off the hoe freely, the mix is ready. If the sides of the furrow collapse, the mix is too wet—add a small amount of the dry materials and retest. If the mortar does not shake freely off the hoe, it is too dry—add very small amounts of water.

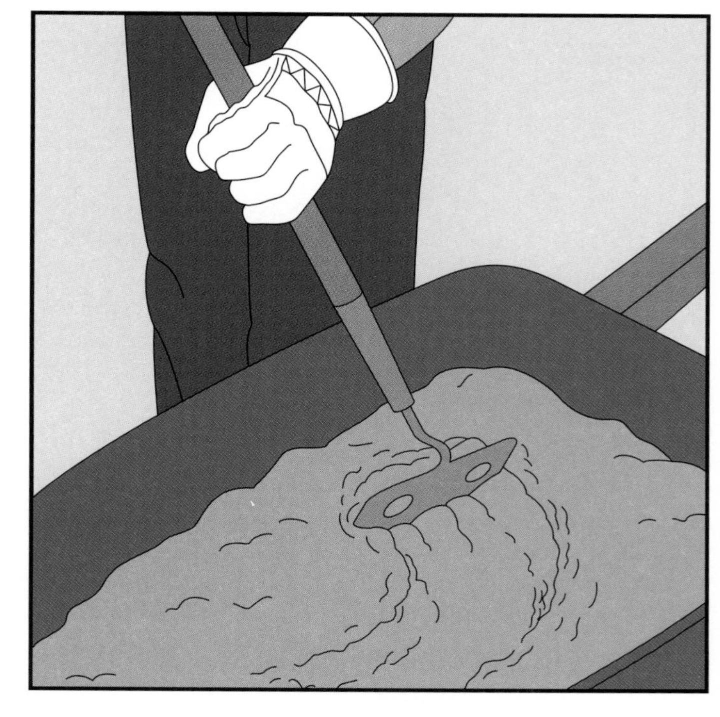

MORTAR RECIPES

Type	Portland Cement	Hydrated Lime	Sand
M	$2\frac{1}{2}$ gal.	$\frac{1}{2}$ gal.	$7\frac{1}{2}$ gal.
N	$1\frac{1}{4}$ gal.	$1\frac{1}{4}$ gal.	$7\frac{1}{2}$ gal.
S	$2\frac{1}{2}$ gal.	$1\frac{1}{4}$ gal.	$8\frac{3}{4}$ gal.

MAKING A MORTAR BED

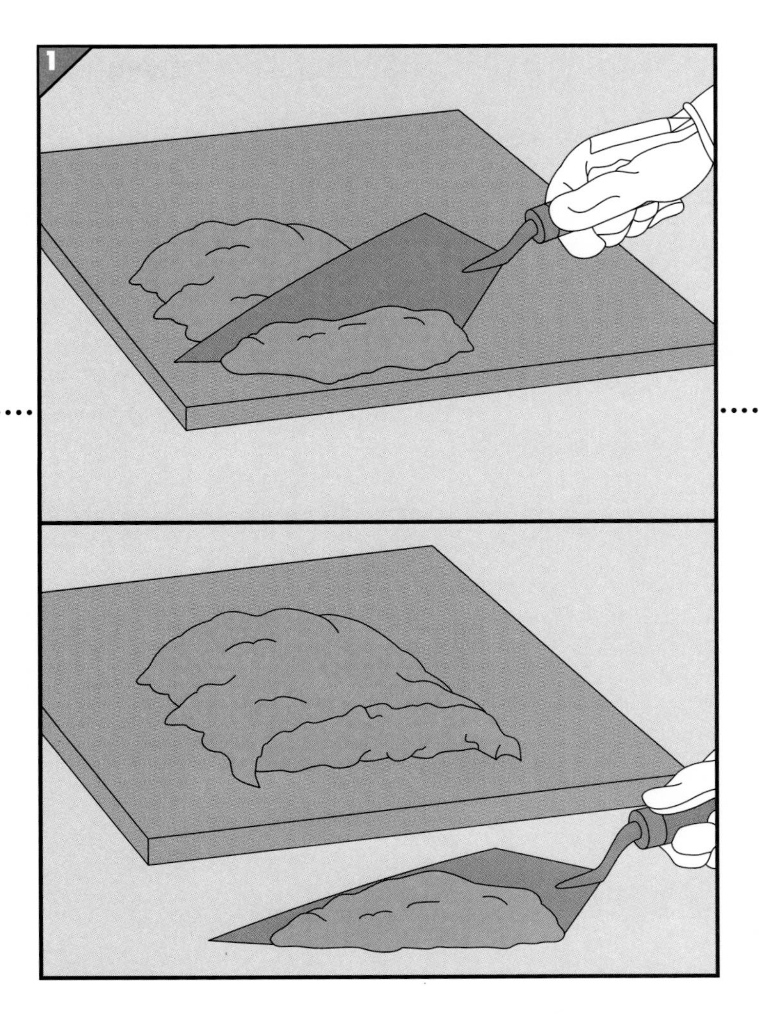

1. Loading the trowel.

◆ With a mason's trowel, scoop mortar from the wheelbarrow and form a mound in the center of a mortar board or a square piece of plywood.

◆ Grasping the trowel handle between thumb and forefinger, drop the edge of the trowel to separate a slice of mortar from the mound *(right, top)*.

◆ With a twist of the wrist, sweep the trowel blade under the slice and scoop up a wedge of mortar onto one side of the blade *(right, bottom)*. Keeping the blade flat, shake the trowel vigorously to flatten the mortar.

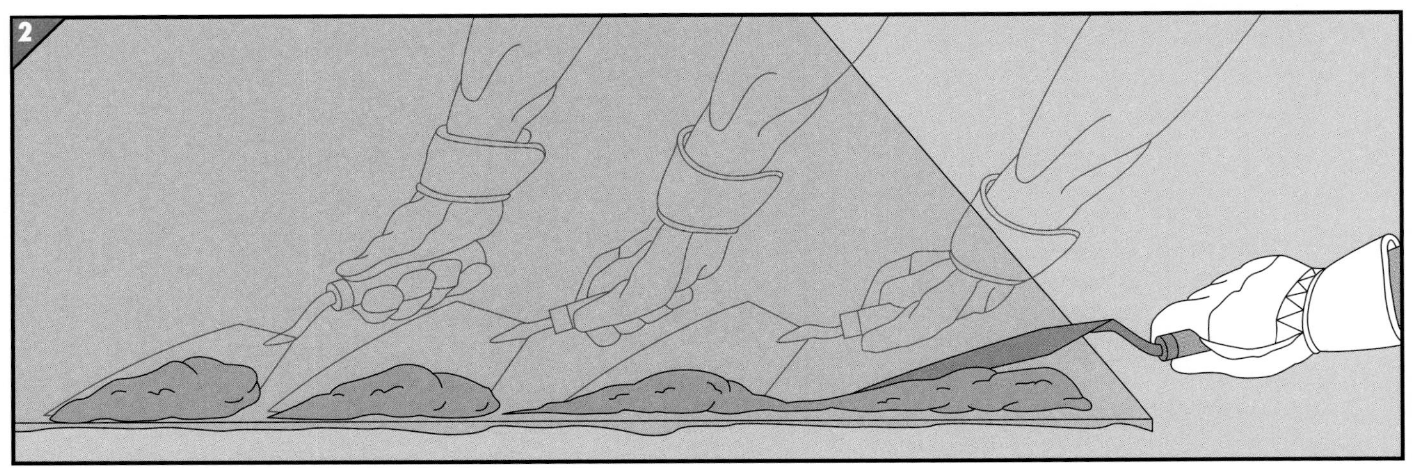

2. Throwing a line of mortar.

You'll need one line—or bed—of mortar to lay a row of bricks; throw two lines for concrete blocks.

◆ Set the point of the trowel at the point where you want to begin the mortar bed.

◆ Pull the trowel toward you and at the same time rotate the blade counterclockwise 180 degrees *(above)*; the mortar will roll off and form a straight line—about one brick wide, a few bricks long, and 1 inch thick. If the line is not straight, return the mortar to the board and try again. Practice until you are able to form a mortar bed three bricks long with one smooth motion.

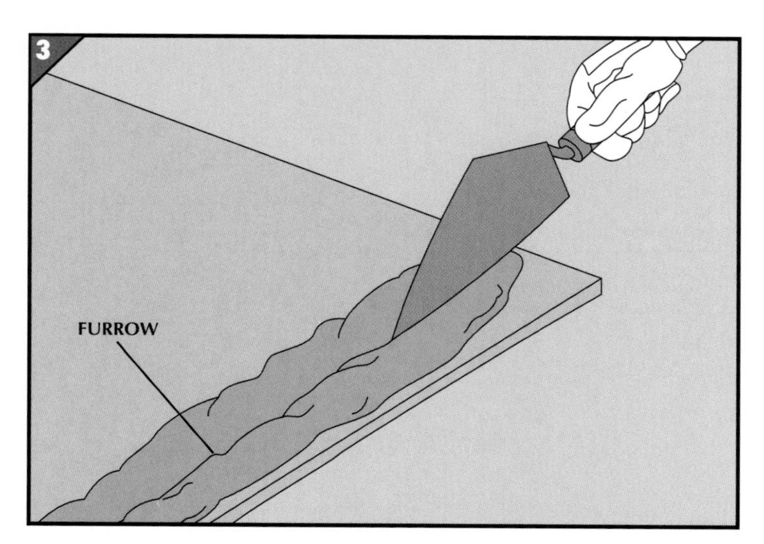

3. Furrowing the mortar.
Immediately after throwing the mortar bed, cut a shallow depression down the center with the point of the trowel *(left)*. By spreading the mortar out slightly from the center, the furrow allows the mortar to be evenly distributed when the brick is pressed down on it.

LAYING BRICKS

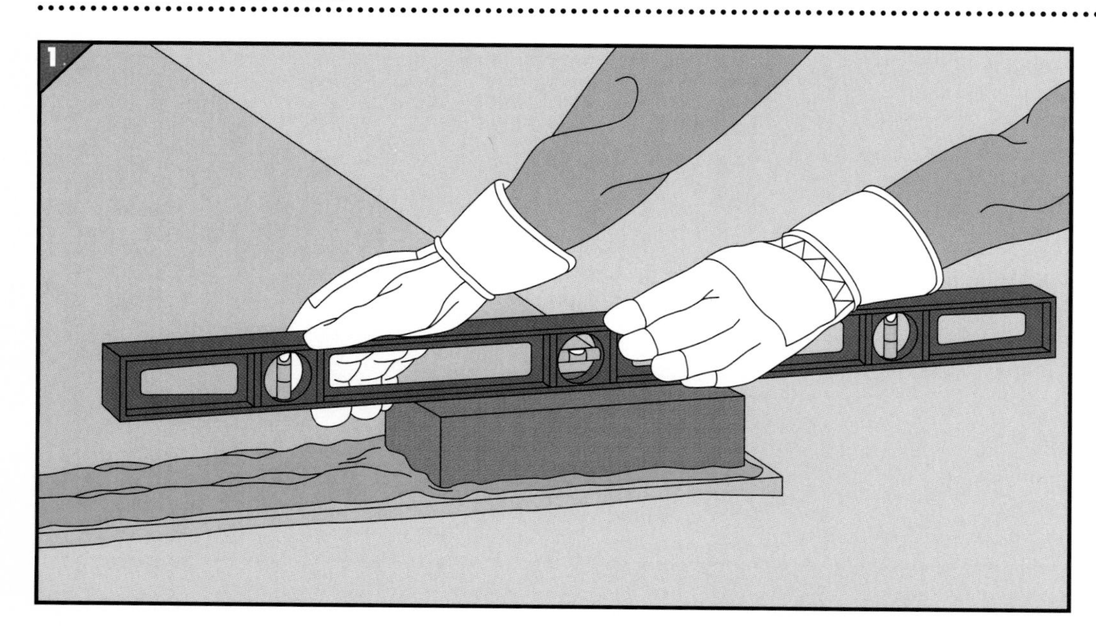

1. Beginning a row.
◆ Starting just inside one end of the mortar bed, push the brick down about $\frac{1}{2}$ inch into the mortar.
◆ With a mason's level—or a long carpenter's level—check that the brick is level—both across its width and length *(left)*—and plumb. If it's not, tap it with the trowel handle and check again. You may need to remove the brick and reset it in the bed to get it in the correct position.

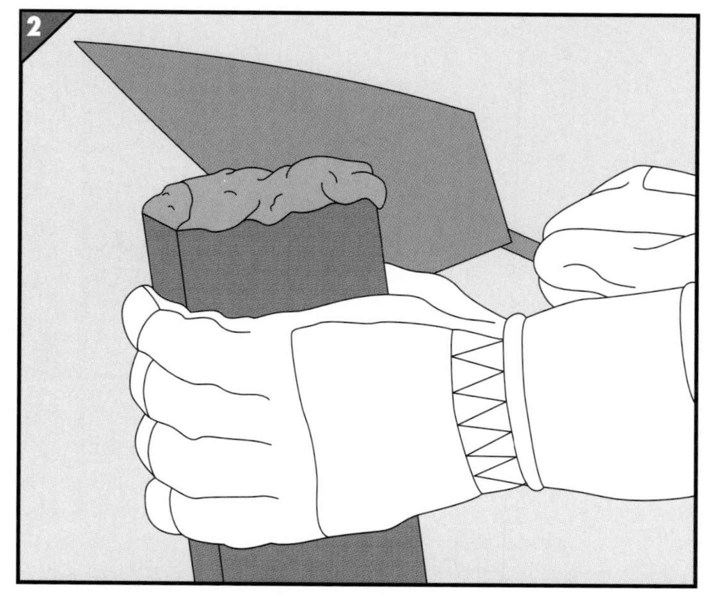

2. Buttering successive bricks.
For all bricks between the ends of a row, spread—or butter—$\frac{3}{4}$ inch of mortar on any surface that will adjoin previously laid bricks. Scoop up enough mortar to cover the surface and spread it on. For end-to-end bricks, cover an end *(right)*; for side-to-side bricks, cover one side. Remove any mortar that slides onto an adjoining surface.

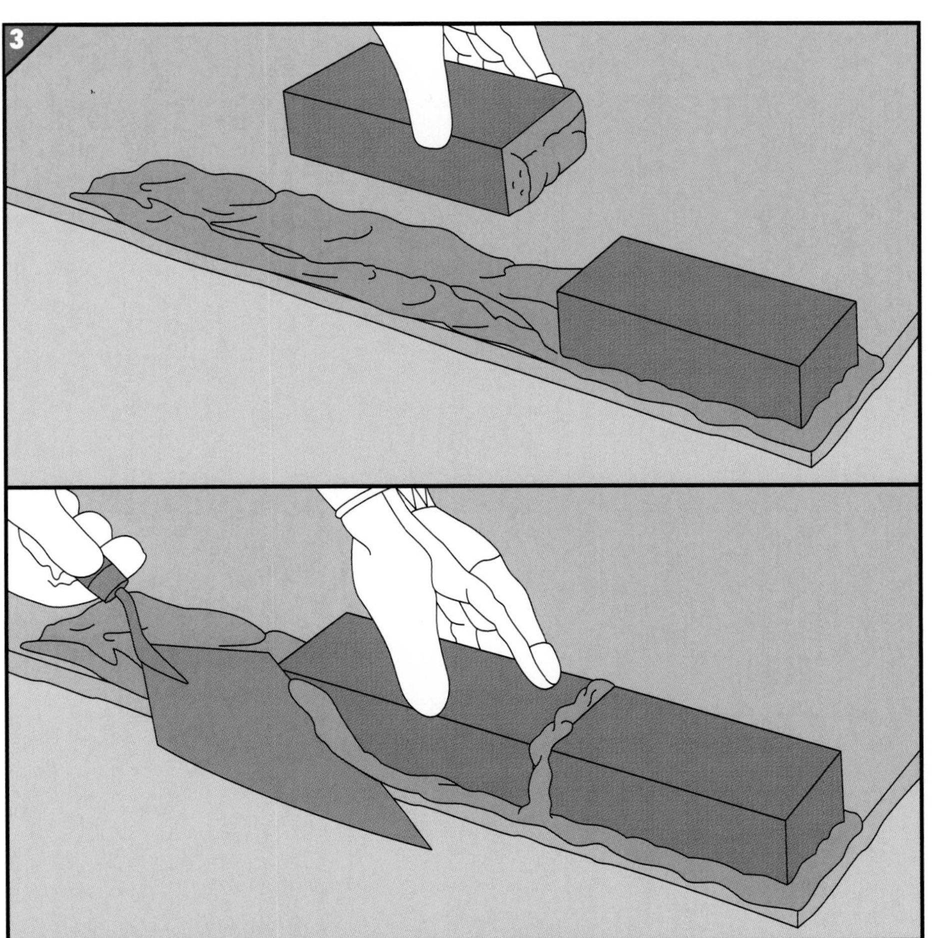

3. Continuing the row.

◆ Set the brick in the mortar bed with the buttered end aligned with the end of the first brick *(left, top)*.

◆ With one motion, push the brick down $\frac{1}{2}$ inch into the bed and against the first brick so there is $\frac{1}{2}$ inch of mortar between them.

◆ With the edge of the trowel, trim off mortar that squeezed out of the joints *(left, bottom)* and return it to the mortar board.

◆ Continue laying bricks until you reach the middle of the row, as determined by the dry run when planning the layout *(page 81)*, then work from the opposite end until there is room for a final brick in the middle.

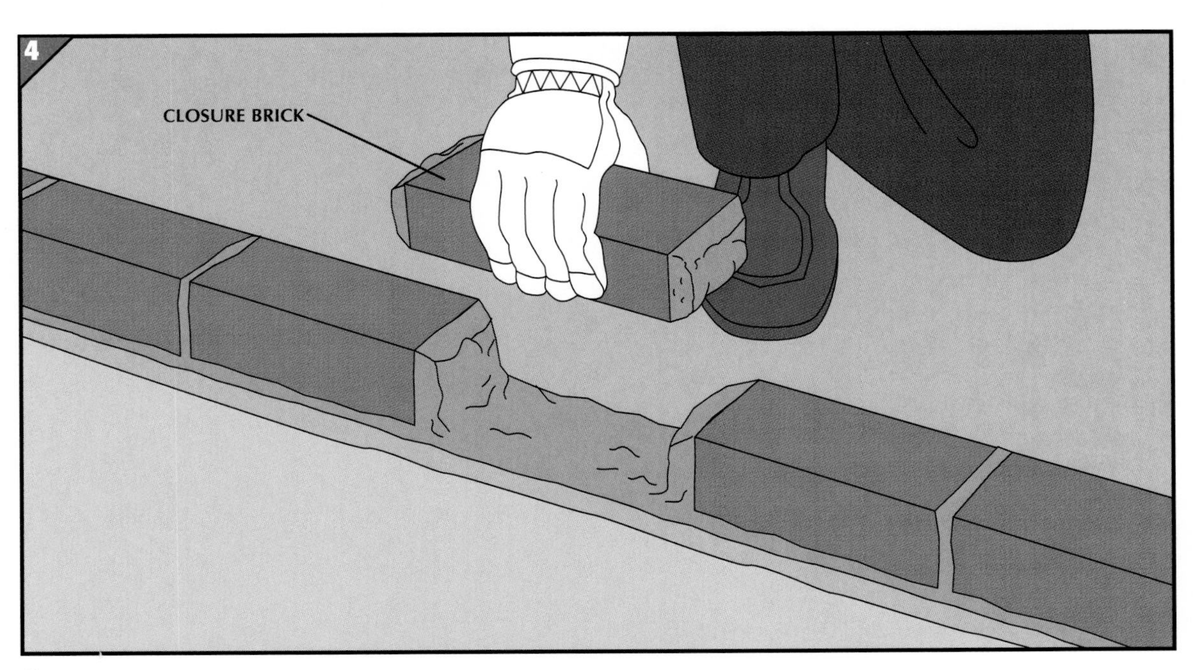

4. Laying the closure brick.

◆ Butter the exposed ends of the bricks on each side of the middle of the row and both ends of the last brick—called the closure brick.

◆ Gently lower the closure brick into the opening *(above)*.

◆ Push the brick into the mortar bed and trim off excess mortar.

FINISHING JOINTS

Concave joint.
This popular joint keeps moisture out and, since mortar is forced tightly between the bricks, makes an excellent bond. Shape it by pressing the mortar firmly with anything curved that fits between the bricks—a convex jointer *(photograph)*, a dowel, a metal rod, or even a spoon.

V-joint.
Form the sharp, water-shedding line of this mortar joint with a V-shaped jointer *(photograph)*, a piece of wood, or the tip of a pointing trowel.

Raked joint.
With its deep recess, this joint is not water-resistant, but it casts a dramatic shadow that accentuates the rows of bricks. Shape the joint by removing $\frac{1}{4}$ inch of mortar and smoothing the surface with a raking tool—a wheeled type can also be used *(photograph)*.

Weathered joint.
This is a water-shedding joint that is recessed from bottom to top. Hold the blade of a pointing trowel at an angle and compress the mortar starting from the front edge of the brick below the joint upward to a point $\frac{1}{4}$ inch inside the brick above.

Struck joint.
Although this joint is not water-resistant because its recess slants from top to bottom, it produces dramatic shadows. To shape it, hold the blade of a pointing trowel at an angle and compress the mortar from the front edge of the top brick to a point $\frac{1}{4}$ inch inside the front of the bottom brick.

Flush joint.
The easiest of all joints to form, this type is neither strong nor water-shedding because the mortar is not compacted. It is made by simply troweling off excess mortar flush with the surface of the bricks.

Extruded joint.
Also called a weeping joint, an extruded joint gives a rustic appearance to a wall. However, it is a poor choice for areas exposed to strong winds, heavy rain, or freezing temperatures because the mortar is not compacted. Create extruded joints by applying excess mortar; when the bricks are laid, some mortar squeezes out and hangs down. You can reproduce the effect when repointing by adding mortar along the joints *(left)*.

Repairs to Brick

For a masonry structure to survive—especially in cold regions—cracks and crumbled sections must be repaired promptly. Water that infiltrates the brick and mortar can cause damage as it freezes and thaws. The techniques shown here for repairing brick apply to paths and patios as well as walls—only the orientation of the work is different.

Mortar Joints and Bricks: Crumbling mortar joints are repaired by a process called pointing—chiseling out the old mortar and packing in new *(below and opposite)*. For damaged bricks, replace the bricks and mortar *(page 18)*; finish the joints to match existing ones *(pages 14-15)*. To piece in a new brick, you may need to cut it to fit *(opposite)*.

Cracks in Walls: Closely inspect any crack in a brick wall—it can be a sign of foundation movement. A crack is usually not serious if its edges are no more than $\frac{1}{8}$ inch apart, parallel, and aligned. Such cracks can be filled using the pointing technique; you may want to color the mortar to match the brick. If a crack is wider than $\frac{1}{8}$ inch, or its edges are misaligned (not matched in shape or position), serious structural problems may exist—consult a building professional.

TOOLS

Joint chisel
Ball-peen hammer
Hawk
Pointing trowel
Joint filler
Ruler
Brick set

MATERIALS

Mortar ingredients:
Portland cement,
masonry sand,
hydrated lime
Replacement bricks

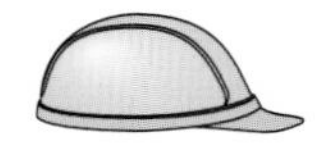

SAFETY TIPS

Dry or wet, mortar is caustic, so protect your hands with gloves, and put on a dust mask when mixing mortar. Always wear goggles when mixing or chipping out mortar and when splitting bricks. Wear gloves when splitting or handling rough bricks.

RENEWING DETERIORATING MORTAR

1. Cleaning out the joint.
◆ With a joint chisel (or a cold chisel) and a ball-peen hammer, chip out crumbling mortar from the joints to a depth of at least 1 inch *(right)*.
◆ Brush or blow the joints clean.

2. Packing in new mortar.

◆ Dampen the joints with a wet brush or a garden hose set to a fine spray.

◆ Spread a $\frac{1}{2}$-inch-thick mound of mortar onto a hawk.

◆ For a short joint, slice off a thin wedge of mortar with the bottom edge of a pointing trowel and press it into the joint *(above, left)*. For long horizontal joints, use a joint filler; align the hawk flush with the opening and push in the mortar *(above, right)*.

◆ When the mortar is firm enough to hold a thumbprint, finish the joints to match the wall *(pages 14-15)*.

SPLITTING A BRICK

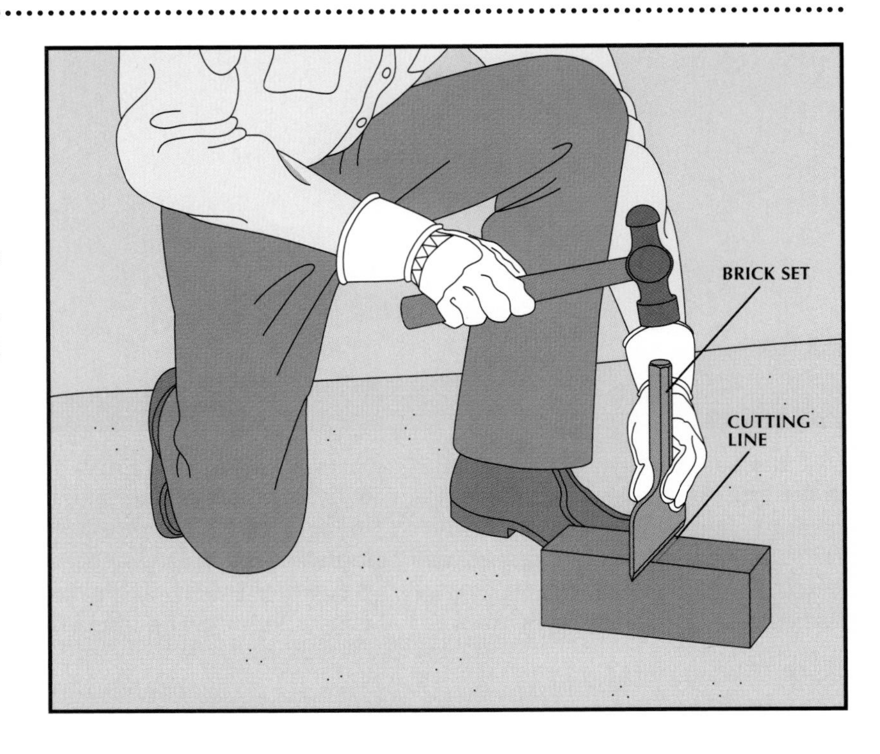

Scoring and cutting.

◆ With a pencil and ruler, mark a cutting line across both side edges of the brick; mark on the diagonal if this shape is required.

◆ Set the blade of a brick set—beveled edge facing the waste portion of the brick—on the cutting line. Tap the handle with a ball-peen hammer to score the line *(right)*.

◆ Repeat to score the cutting line on the opposite edge.

◆ Lay the brick on a bed of sand with the waste part of the brick pointing away from you. Insert the brickset into the scored line, again with the bevel facing the waste, and strike the handle sharply to split the brick.

REPLACING A BROKEN BRICK

Fitting in a new brick.
◆ Chisel out the mortar surrounding the damaged brick *(page 16, Step 1).*
◆ Chip out the brick with a brick set and ball-peen hammer, then brush the space clean.
◆ Select a brick that fits the slot or cut one to fit *(page 17).*
◆ Dampen the slot's surfaces and apply a $\frac{3}{4}$-inch coating of mortar with a pointing trowel.
◆ Hold the brick on a hawk about $\frac{1}{2}$ inch above the row and push the brick into the slot.
◆ Trowel in extra mortar if needed to fill the joints *(page 17, Step 2).*
◆ When the mortar is firm enough to hold a thumbprint, finish the joints to match the wall *(pages 14-15).*

FILLING IN A DAMAGED WALL

Cutting out and replacing bricks.
◆ Remove the mortar surrounding the damaged bricks *(page 16, Step 1)* and chip out the bricks with a brick set and a ball-peen hammer *(above, left).* It may be necessary to remove bricks above the damage as well. Brush away debris.
◆ Dampen all surfaces of the replacement bricks and their openings in the wall.
◆ Lay mortar beds for the bricks, troweling and furrowing the bed as you would for new bricks *(page 11).*
◆ Butter the bricks and lay them in place on the mortar beds *(above, right).*
◆ When the mortar is firm enough to hold a thumbprint, finish the joints to match the wall *(pages 14-15).*

Mending Concrete

Paradoxically, concrete surfaces cannot be repaired with concrete—the coarse gravel aggregate in the new concrete would prevent a strong bond between the patch and the surrounding area. Instead, it's best to use mortar or commercial epoxy or latex patching compounds designed for concrete repairs.

Surface Preparation: Remove damaged concrete and all dirt, debris, and standing water, and keep the area damp for several hours—preferably overnight.

Filling Cracks: For cracks up to $\frac{1}{8}$ inch wide, use a latex or epoxy patching compound. Force it into the crack with a putty knife or a mason's trowel, and smooth it level with the surrounding concrete.

Mend larger cracks with mortar prepared without lime *(below and page 20)*. You can also use a latex or epoxy patching compound and apply it in a similar way.

Spalling: Surfaces on which the concrete has flaked off in thin scales—a condition called spalling—are best patched with an epoxy patching compound *(page 21)*.

Repairing Steps: Large broken pieces of steps can be glued back into place with an epoxy patching compound *(page 22)*. But when sections of the stairs crumble away, you will have to build them up again with mortar *(pages 22-24)*.

Curing: A mortar patch must cure slowly and in the presence of moisture. Let the patch set for about two hours, then cover it with a sheet of plastic. On a horizontal surface, the cover can be secured by bricks or rocks; for vertical surfaces, use tape. Over the next three days, lift the cover daily and sprinkle water on the patch. If a vertical patch cannot be covered conveniently, moisten it twice a day. For latex and epoxy compounds, check the product label for curing instructions.

TOOLS

- Cold chisel
- Ball-peen hammer
- Paintbrush
- Mason's trowel
- Steel trowel
- Sledgehammer
- Wire brush
- Putty knife
- Circular saw
- Stair edger

MATERIALS

- Mortar ingredients: Portland cement, masonry sand
- Epoxy concrete-patching compound
- Scrap lumber
- Bricks

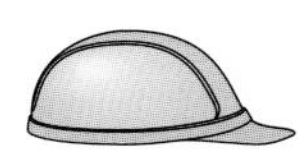

SAFETY TIPS

Wear gloves when repairing concrete, and goggles when breaking up a damaged surface.

FILLING WIDE CRACKS

1. Removing damaged concrete. With a cold chisel and a ball-peen hammer, chip away all cracked or crumbling concrete to about 1 inch below the surface *(right)*.

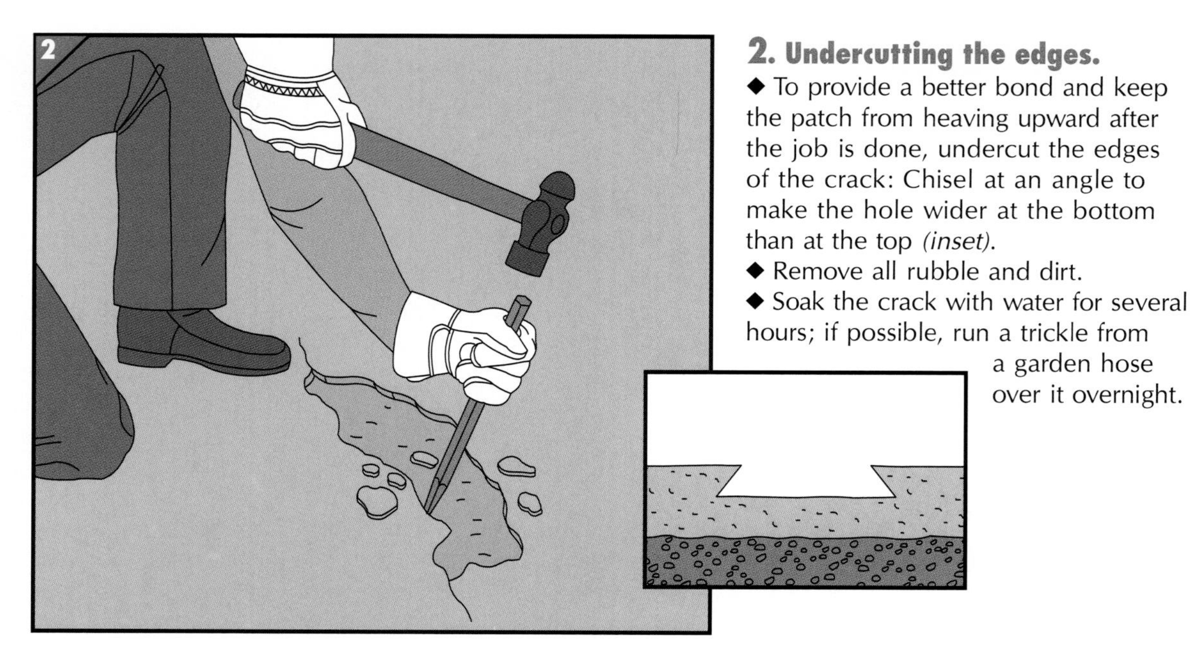

2. Undercutting the edges.

◆ To provide a better bond and keep the patch from heaving upward after the job is done, undercut the edges of the crack: Chisel at an angle to make the hole wider at the bottom than at the top *(inset)*.
◆ Remove all rubble and dirt.
◆ Soak the crack with water for several hours; if possible, run a trickle from a garden hose over it overnight.

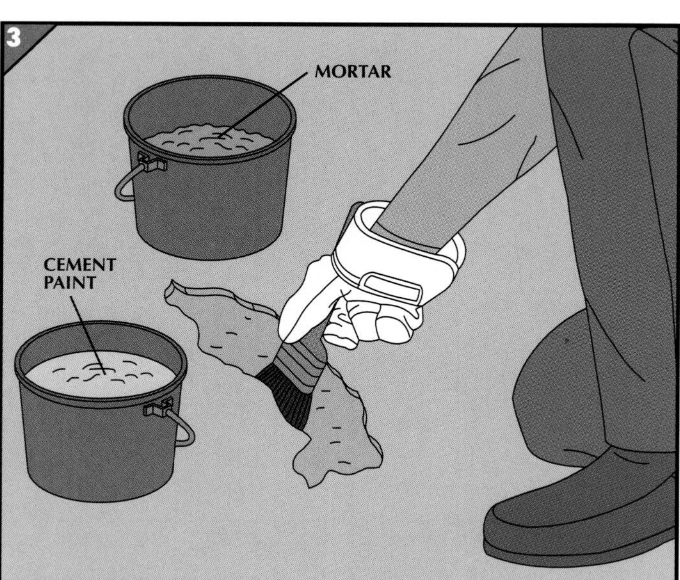

3. Preparing the crack.

◆ Prepare the mortar by mixing 1 part Portland cement and 3 parts masonry sand, adding enough water to make a paste stiff enough to work with a mason's trowel.
◆ Make a small batch of cement paint by adding water to Portland cement until it has the consistency of thick paint.
◆ Coat the edges of the crack with the cement paint *(left)*; then, proceed immediately to Step 4 to complete the repair before the paint dries.

4. Mending the crack.

◆ Pack the mortar firmly into the crack with a mason's trowel, cutting deep into the mixture to remove air pockets.
◆ Level the mortar with a steel trowel.
◆ Let the patch stand for an hour, then spread it evenly across the surface, sliding the trowel back and forth with its leading edge raised *(right)*.

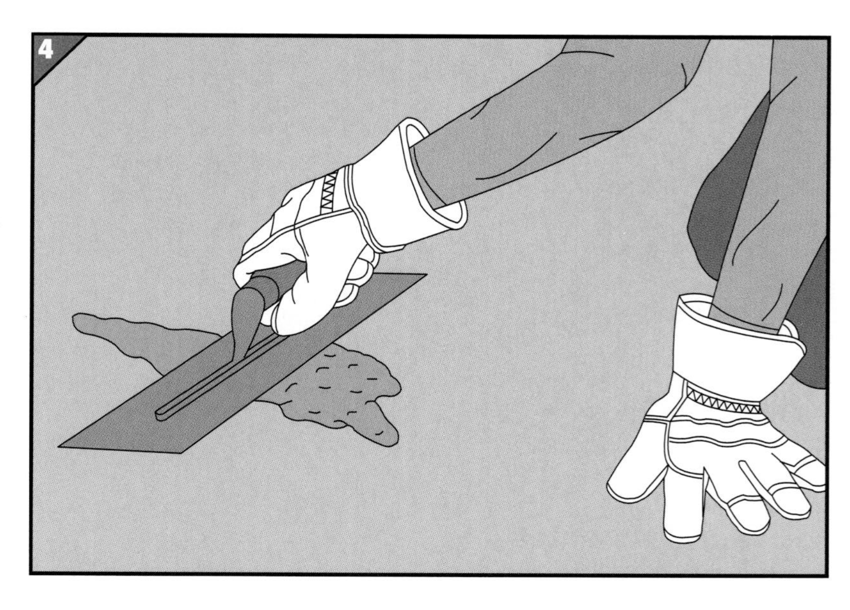

TREATING SPALLED SURFACES

Using an epoxy mix.

◆ Break up large areas of scaling concrete with an 8-pound sledgehammer *(right)*—don't slam the tool against the surface; let its own weight provide the force. A rotary hammer with a bush hammer head *(below)* is an alternate tool for the job; for small areas, a ball-peen hammer and a cold chisel will be adequate.
◆ Sweep up dust and debris; dislodge small fragments with a stiff wire brush.
◆ Soak the damaged area with water and keep it wet for several hours, preferably overnight. The area should still be damp when you apply the patch.
◆ Prepare a commercial epoxy patching compound for concrete and apply it with a steel trowel. Bring the new layer level with the surrounding concrete, and feather it thinly at the edges.
◆ Let the patch stand for 24 hours before letting it support any weight.

A POWER TOOL TO BREAK UP CONCRETE

A rotary hammer—available at tool rental shops—produces a rapid chiseling action similar to a jackhammer; a side handle provides a sturdy grip. Rotary hammers come with a variety of heads—for a concrete slab, choose a bush hammer head, as shown. Since the tool is relatively small, you may prefer a jackhammer equipped with a bush hammer head for a large job.

A rotary hammer is also handy for boring holes in masonry—replace the bush hammer head with a masonry core bit, similar to a hole saw—and a flip of a switch provides the drilling action.

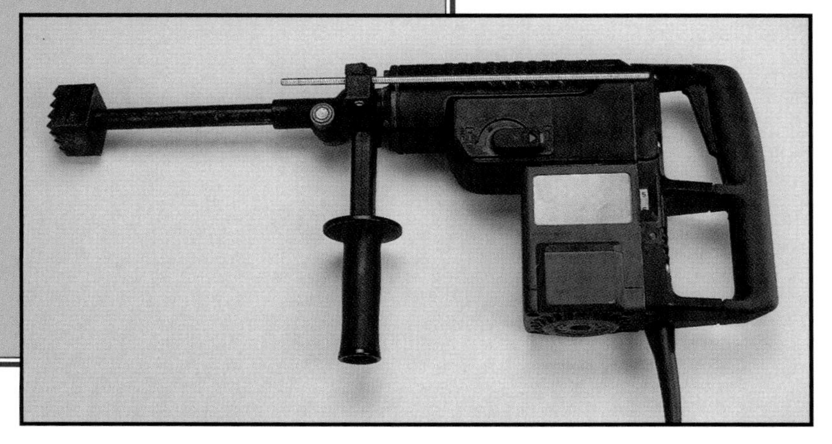

GLUING A BROKEN STEP

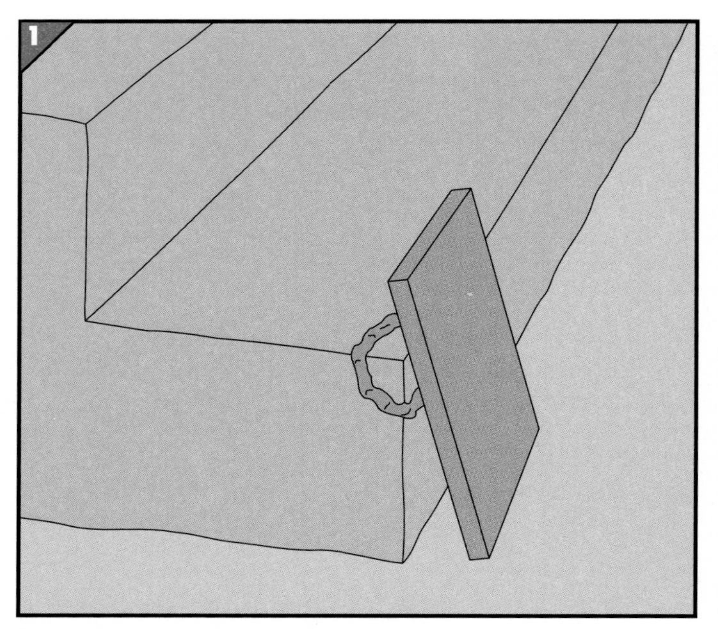

1. Gluing the chip.
◆ Brush particles of dirt and cement from the broken piece and the corner of the step.
◆ Mix a small batch of epoxy patching compound for concrete; then, with a mason's trowel, spread some onto the chipped part of the broken piece.
◆ Hold the piece firmly in place until the compound hardens; you can prop a board against the piece to hold it in place *(above).*

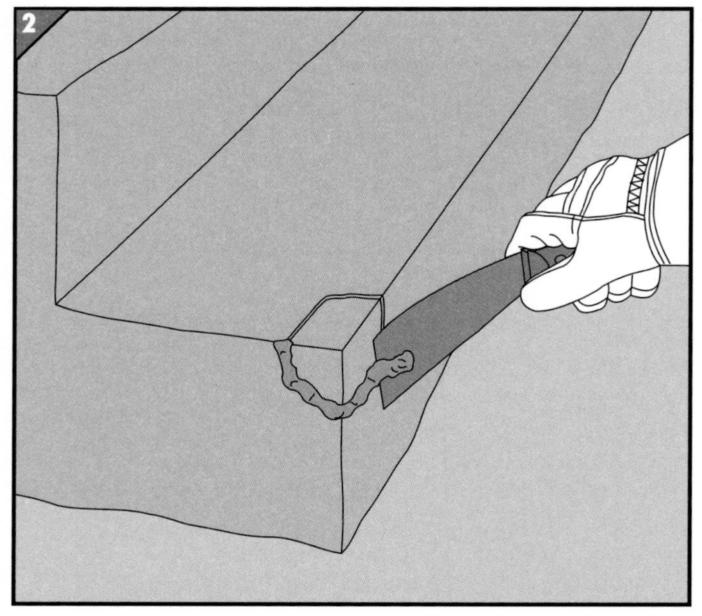

2. Completing the job.
◆ Once the compound has set, use a putty knife to scrape away any excess that has oozed out between the piece and the step *(above).*
◆ If a small crack remains around the repair, pack patching compound into the crack with a trowel, then smooth the patch level.
◆ Avoid touching the repaired corner for at least 24 hours.

REBUILDING A CORNER

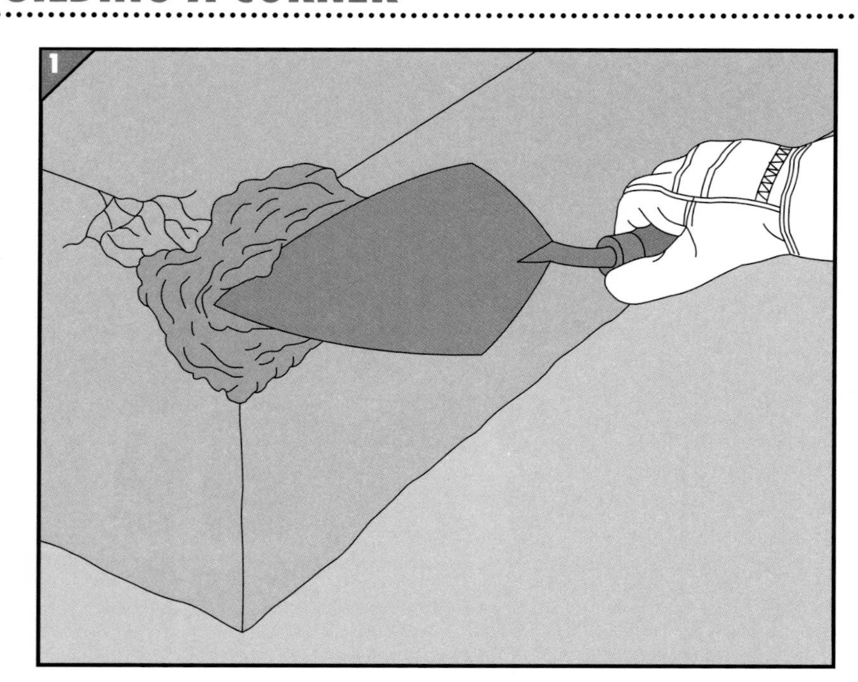

1. Shaping a replacement piece.
◆ Clean the corner and keep it damp for several hours, preferably overnight.
◆ Mix 1 part Portland cement with 3 parts masonry sand and add just enough water to make a mortar paste that holds its shape.
◆ With a mason's trowel, apply the mortar to the damage, roughly shaping it. Let the patch harden until it is firm enough to hold a thumbprint.

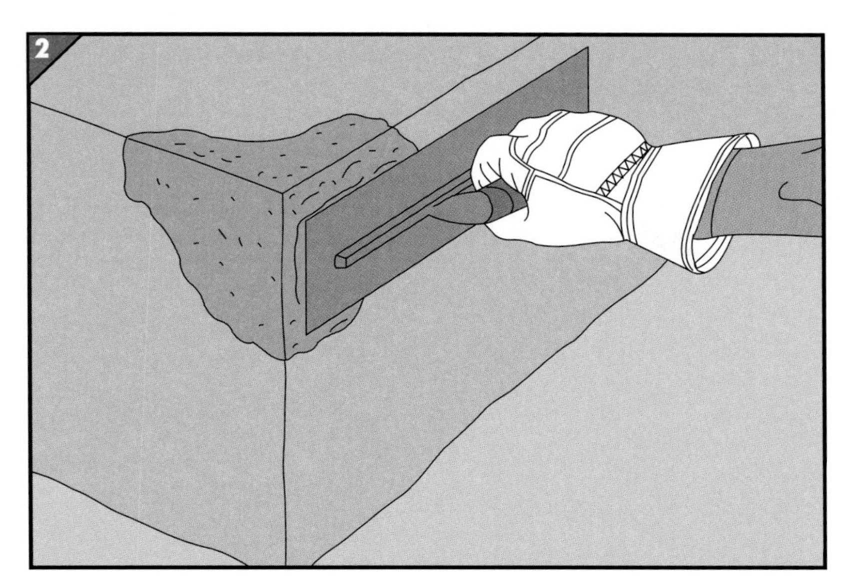

2. Finishing the corner.
◆ Finish and smooth the corner flush with the steps with a steel trowel *(left)*.
◆ Let the mortar cure for at least three days, and avoid putting weight on the corner for a few days afterward.

REPAIRING A CHIPPED EDGE

1. Clearing the damage.
With a cold chisel and a ball-peen hammer held horizontally, chip off the damaged concrete all the way across the edge of the step.

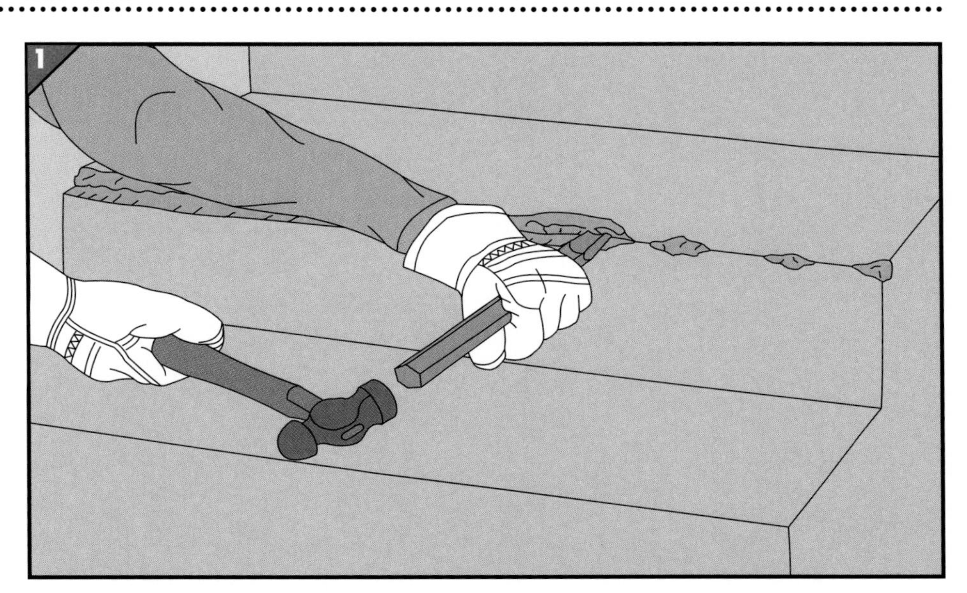

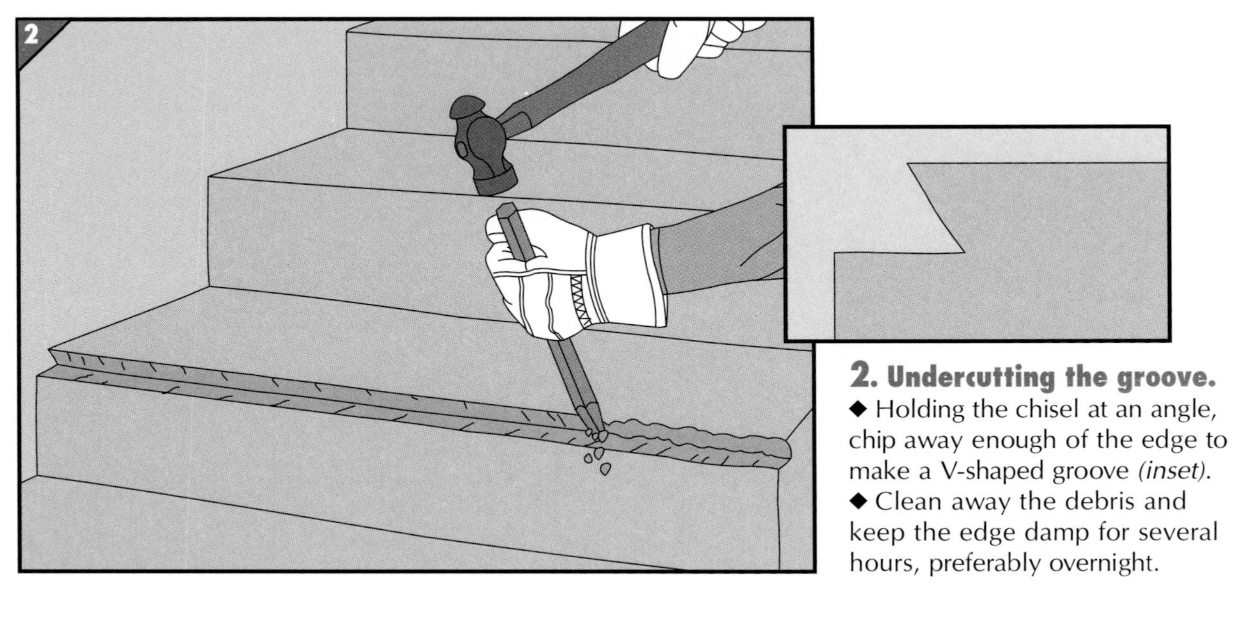

2. Undercutting the groove.
◆ Holding the chisel at an angle, chip away enough of the edge to make a V-shaped groove *(inset)*.
◆ Clean away the debris and keep the edge damp for several hours, preferably overnight.

3. Rebuilding the edge.

◆ Cut a board to the length and height of the riser and set it against the step as a form board. Hold it in place with bricks or concrete blocks.

◆ Mix 1 part Portland cement with 3 parts masonry sand, then add just enough water to make a mortar paste that holds its shape.

◆ Coat the groove with cement paint *(page 20, Step 3)*, then immediately fill it in with the mortar, using a steel trowel to shape it and smooth it flush with the step and form board *(right)*.

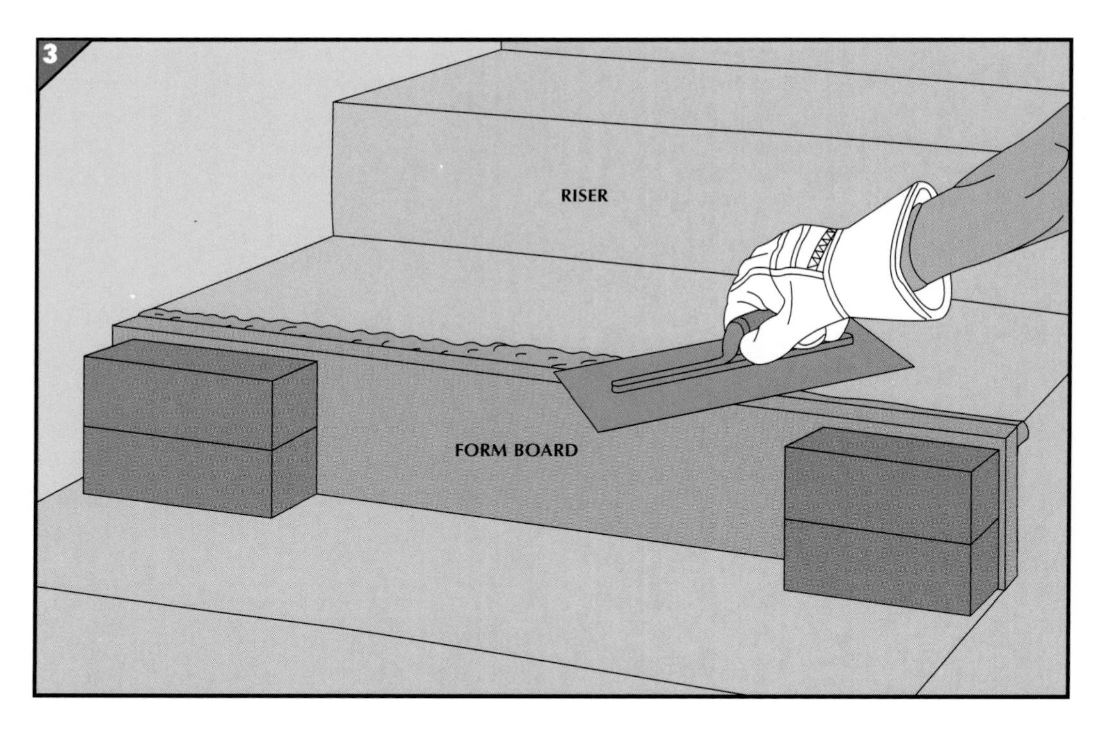

4. Finishing the job.

Once the mortar is thumbprint hard, round the edge of the step with an edger *(above)*, then carefully remove the form board. Let the mortar cure for at least three days, then avoid stepping on the edge for a few more days.

First Aid for Holes in Asphalt

The asphalt used to cover driveways and walks is a mixture of gravel with a crude-oil extract as a binder. This material can develop cracks and holes from frost, water, and traffic; it also absorbs oil and other automotive drippings.

Patching Holes: A type of filler called cold-mix asphalt is manufactured in two varieties: cut-back and emulsified. Either product works well in dry holes, but damp holes require the emulsified type. Both types are sold in airtight bags or buckets, usually containing enough material to patch about $1\frac{1}{2}$ square feet of surface.

When the temperature is below 40°F, do not patch or seal asphalt. If cool weather has hardened the cold-mix into an unworkable lump, soften it by placing the container in a heated area for a few hours before use.

Sealing the Surface: To protect asphalt from deterioration and from absorbing oil and automobile fluids, coat it once every four or five years with a waterproof and petroleum-resistant sealer containing sand for skid resistance. Simply pour the product onto the surface and spread it evenly with a push broom or squeegee.

The sealer will fill cracks up to $\frac{1}{8}$ inch wide. For cracks up to $\frac{1}{2}$ inch wide, clean out soil and debris, then pour ready-to-use crack filler into the cavities. If a crack is up to 1 inch wide, thicken the filler with sand to a puttylike consistency, then push the mixture into the crack with a putty knife.

TOOLS

Shovel
Circular saw and masonry blade

MATERIALS

Cold-mix asphalt
Sand
4 x 4
Door handles
Wood screws

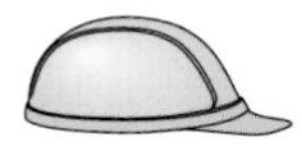

SAFETY TIPS

Wear gloves when working with cold-mix asphalt.

USING COLD MIXES

1. Preparing the hole.

◆ With a shovel, dig out the hole to a depth of 3 or 4 inches and remove any loose material.
◆ Cut back the edges of the hole with a circular saw and a masonry blade until you reach sound asphalt, making the sides of the hole vertical.
◆ Compact the bottom of the hole with a tamper made by screwing a pair of large door handles to opposite sides of a 4-by-4 *(left)*.

2. Patching the hole.

◆ Pour cold-mix asphalt into the hole, filling it halfway.
◆ Slice through the asphalt with a shovel to release air pockets *(left)*, then compact it with the tamper.
◆ Complete the patch by adding 2-inch layers of cold-mix, tamping each one, until it is even with the surrounding surface. On a driveway, build up the patch $\frac{1}{2}$ inch higher than the surface, since the weight of a car will flatten it.
◆ Spread sand over the patch until it dries (usually about two days) to keep it from sticking to footwear.

Homemade Concrete in Convenient Batches

Concrete's great strength comes from its materials: gravel (called coarse aggregate), sand, and cement. The coarse aggregate supplies bulk, the sand fills voids between the aggregate, and the cement, when moistened with water, binds the sand and aggregate into a durable solid.

Air-entrained Concrete: For most projects, you can use ordinary concrete. But in an area subject to freezing, large projects, such as building a concrete patio or steps, require air-entrained concrete, which contains additives that produce and trap microscopic air bubbles. When the concrete dries, the bubbles form tiny spaces within the slab so it can expand and contract with a minimum of cracking.

Mixing Concrete: It's best to buy sacks of premixed dry ingredients to which you simply add water. One 60-pound bag will make about $\frac{1}{2}$ cubic foot of concrete. The amount of water required for a concrete recipe is critical; even a small amount of extra water can weaken the concrete. The approximate amount of water will be indicated on the bag, but add it only a little at a time, and be sure to test the consistency as described on page 28.

A small amount of plain concrete (about $2\frac{1}{2}$ cubic feet) can be mixed by hand in a wheelbarrow *(opposite)* or on any flat surface. Unlike ordinary concrete, the air-entrained type must be machine mixed; buy a small quantity of the additive from a concrete supplier and rent a gasoline- or electric-powered mixer *(opposite, bottom)*. The mixer will enable you to pour and finish about 16 square feet of a slab 4 inches thick before the concrete becomes too hard to finish. If you need more, arrange for a truck to deliver ready-mix concrete *(page 37)*.

TOOLS

Mortar hoe
Square shovel
Bucket
Wheelbarrow

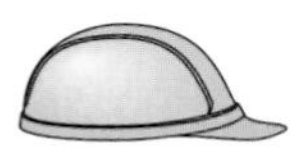

MATERIALS

Premixed concrete

SAFETY TIPS

Wet or dry, concrete is caustic—wear gloves, goggles, and a dust mask when mixing concrete.

AN EXTRA INGREDIENT: COLOR

Concrete, though normally gray, can be colored by several methods. You can paint or stain the concrete after it cures, but it will have to be repainted as the color wears away. You can also apply a dust-on pigment while the concrete is still wet. The color is applied in two coats and the surface is floated before and after each application *(page 50, Step 3)*. This pigment is inexpensive but can be tricky to apply evenly. An alternative to these methods is to add pigment to the concrete mixture. Although more expensive than the other techniques, the color is distributed evenly and is permanent. If you order ready-mix concrete, you can request that it be precolored.

Adding color to concrete requires machine mixing; simply add synthetic pigments or natural metallic oxides to the dry concrete in the power mixer. Synthetic pigments cost more, but are less likely to bleach or fade; both types come in a range of colors you can use directly or combine to produce custom shades. To keep coloring uniform from batch to batch, measure all ingredients by weight, not volume; and never add more than 10 percent pigment—it will weaken the concrete. A bathroom scale wrapped in plastic is ideal for weighing. Concrete lightens as it dries, so you may need to experiment to find the right amount of pigment.

PREPARING CONCRETE

Mixing in a wheelbarrow.

◆ Empty the premixed dry ingredients into a wheelbarrow. With a mortar hoe, push the mixture to the sides to form a bowl-like depression.
◆ Slowly pour about three quarters of the required quantity of water into the depression.
◆ Pull the dry materials from the edges of the ring into the water, working all around the pile until the water is absorbed by the mixture. When no water remains standing on the surface, turn the concrete over with the hoe three or four times.
◆ Add water a little at a time until the mixture completely coats all the coarse aggregate. Leave any unused water in the bucket until you test the consistency of the concrete, then make any necessary corrections *(page 28)*.

MIXING BY MACHINE

If you're mixing more than 3 cubic feet of concrete, it's worth renting an electric or gas-powered mixer. A typical electric mixer holds 6 cubic feet, but they are available in capacities up to 12 cubic feet.

Set up the mixer near your work site and make sure it is level and wedged in place.

To make concrete from premixed dry ingredients, empty the dry ingredients and about half the water into the mixer. Turn on the mixer for about three minutes. Then add air-entrainment additive, if needed, and gradually add more water until the mixture completely coats the coarse aggregate and the concrete is a uniform color.

Test a few shovelfuls in a wheelbarrow *(page 28)*, and return the test batch to the mixer before making corrections. When the concrete is thoroughly mixed, dump the contents into a wheelbarrow and hose out the drum. When you finish using the mixer, clean up as described on page 28.

CAUTION *Never reach into the mixer or insert tools while it is turning. Never operate an electric mixer in damp conditions, and cover it when not in use. Fuel a gas-powered mixer only when the engine is off and has cooled down.*

TESTING THE CONSISTENCY

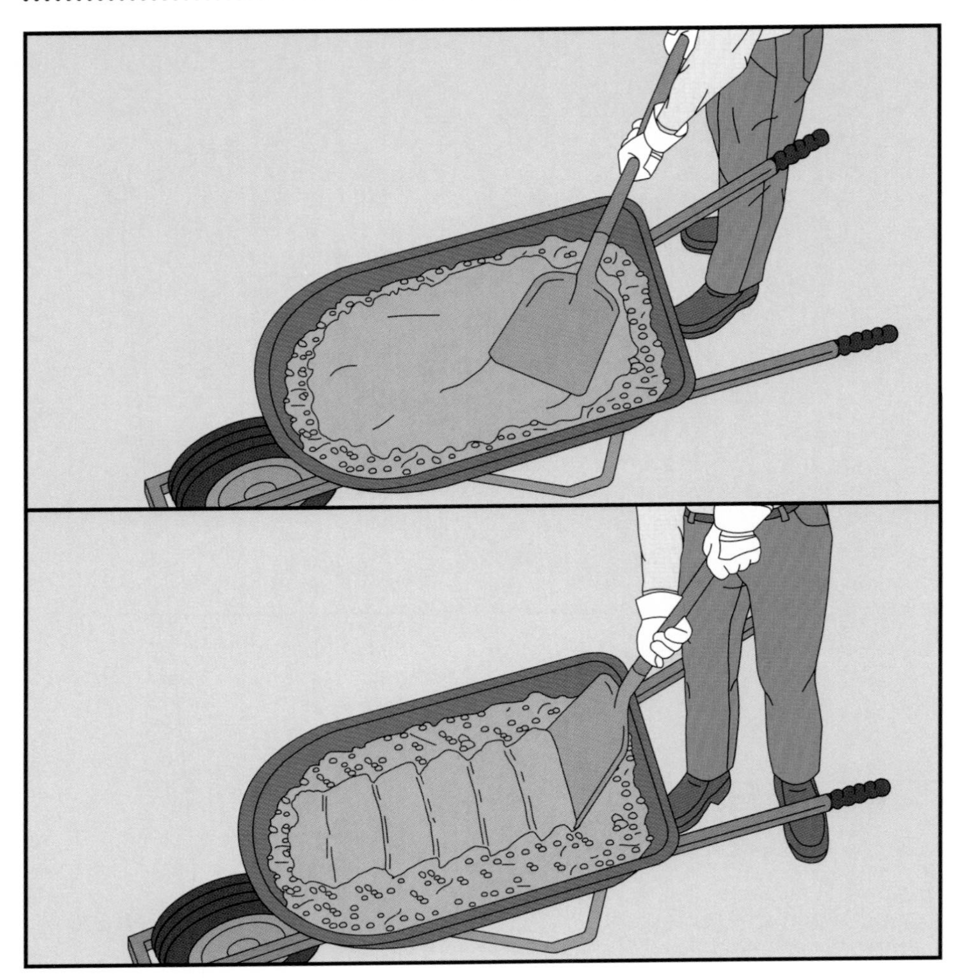

Judging and correcting the mix.

◆ Smooth the concrete in the wheelbarrow by sliding the bottom of a square shovel across the concrete's surface *(left, top)*.

◆ Jab the edge of the shovel into the concrete to form grooves. If the surface is smooth and the grooves are distinct, the concrete is ready to use *(left, bottom)*. If the surface roughens or the grooves are indistinct, add a small amount of water. If the surface is wet or the grooves collapse, add a small amount of dry ingredients.

◆ Retest the batch until the consistency is correct.

CLEANING UP AFTER THE WORK IS DONE

Most sanitation departments will not haul away leftover concrete or mortar. Generally, you have to take it yourself to the nearest dump. For easier handling, pour it into paper bags, or pile it in small heaps on sheets of paper and let it set into manageable lumps. You can also mold concrete and save it for future use—for example, keep simple 2-by-4 wood forms ready for pouring excess concrete while it is still workable, and cast stepping-stones.

Clean tools at the end of each work session. Put all tools in a wheelbarrow and hose them down. Do not dump dirty water into your drainage system or into a city street or sewer. Instead, dig a large hole, pour the dirty water in, and fill the hole.

Hose out a power concrete mixer at day's end. (Many rental companies charge an extra fee for a mixer that comes back dirty.) If you cannot clean the drum completely with a hose, turn the mixer on and pour in a mixture of water and two shovelfuls of gravel to scour it out. Empty the mixer after three or four minutes, then hose it out again. If you have waited too long to clean the mixer, you may have to scrape bits of hardened concrete out with a wire brush or chip them off with a chisel.

Removing Stains and Blemishes

Ordinary scrubbing with detergent and a stiff fiber brush gets most blemishes off brick and concrete; if this method fails, they may need special treatments with chemicals available from a pharmacy, home-improvement or garden center, or a pool-supply store *(below)*. Moss can usually be removed from brick and concrete with ammonium sulfamate. Wash slate, granite, or bluestone with a gentle laundry detergent. For limestone and sandstone, avoid detergents; instead, use clean water and a scrub brush.

CAUTION *Always add acid to water, never water to acid.*

TOOLS

Stiff fiber brush
Paint scraper

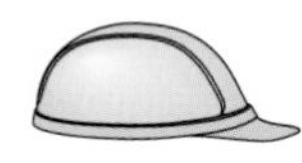

SAFETY TIPS

When working with chemicals, wear goggles and rubber gloves.

Efflorescence and smoke.

The white, powdery deposit known as efflorescence can be cleaned from brick with a commercial product formulated for brick. On concrete or block, as a last resort, apply a 1-to-10 solution of muriatic acid, but don't leave it on the surface more than a few minutes. For colored concrete, dilute the solution to 1-to-100.

Scrub off smoke residue with a scouring powder containing bleach, then rinse with water.

EFFLORESCENCE

SMOKE

Oil, tar, and stains.

Apply a commercial emulsifying agent to oil or tar on brick, then hose it off with water. For stubborn tar, add kerosene to the agent. On concrete or block, apply a degreaser, available at auto-supply stores. Let it stand for the time recommended, then wipe it off.

For brown stains caused by the manganese used for coloring brick, wet the brick and brush it with a solution of 1 part vinegar, 1 part hydrogen peroxide, and 6 parts water. For green stains caused by the vanadium salts in the brick, use a solution of 1 pound of potassium or sodium hydroxide and 1 gallon of water; leave it on for two or three days, then rinse it off. Or use a commercial product formulated for either type of stain.

OIL

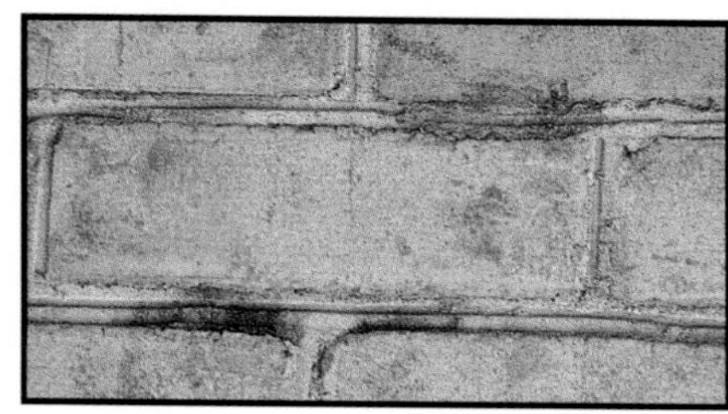

BROWN STAIN

Paint and rust.

To remove paint, apply a commercial water-based paint remover; clean it off with a scraper or a wire brush, then wash the surface with water.

To get rust off brick, spray or brush the brick with a strong (1 pound per gallon of water) solution of oxalic acid, then hose it off. On concrete or block, scrub it with a stiff brush and a solution of 1 part sodium citrate, 7 parts glycerin, and 6 parts lukewarm water; then rinse it thoroughly.

PAINT

RUST

Fastening to Masonry

Many home improvement projects, such as putting up a porch railing, installing outdoor wiring, or running a water pipe to an outside faucet, involve anchoring to masonry. The fastener you choose depends partly on the object you are mounting and partly on the type of masonry. Screws and expansion anchors, toggle bolts, steel masonry nails, and a family of glues called mastics are the principal types of fasteners.

Masonry Nails: For relatively light loads such as furring strips for paneling, use masonry nails. The nails can be driven by hand or by a powder-actuated hammer *(opposite)*, which uses gunpowder to fire the nails. Some are activated by a trigger; others must be struck with a hammer. Follow the manufacturer's directions exactly when using this tool. Brick and stone are too hard for nailing into except at mortar joints.

Mastics: These glues, which are simply spread on the masonry surface, give added strength to masonry nails. They can be used on concrete or block, but brick and stone make poor surfaces for gluing.

Expansion Anchors and Toggles: Expansion anchors—consisting of a screw and an anchor—are suitable for concrete and the solid parts of blocks. They can be fastened into mortar joints and in brick, if you are careful not to tighten the screw so much that the material around the edge of the hole begins to crumble. In stone, expansion anchors are unsatisfactory: They can create stresses that will cause cracks. For this material—and any solid masonry where a very strong fastener is required—a technique employing bolts is useful *(opposite)*. Toggles—consisting of a bolt and threaded wings—are suited best for the hollow parts of concrete blocks.

Making the Holes: Toggles and anchors fit into predrilled holes. Small holes can be made in concrete block with a carbide-tipped masonry bit in an electric drill. Holes larger than 1 inch need a four-edged chisel called a star drill. For small holes in harder materials such as concrete, brick, or stone, consider renting an electric hammer drill *(below)*. Holes larger than 1 inch in hard materials need a rotary hammer with a masonry core bit *(page 21)*.

TOOLS

- Carpenter's level
- Utility knife
- Putty knife
- Rubber mallet
- Powder-actuated hammer
- Electric drill

MATERIALS

- Masking tape
- Bolts, washers, nuts
- Epoxy
- Power fasteners

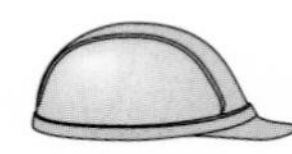

SAFETY TIPS

Always wear goggles when drilling or nailing into masonry; add earplugs when using a powder-actuated hammer.

A HAMMER AND DRILL ALL IN ONE

To make small holes in solid masonry—brick, stone, or concrete—you may want to rent a hammer drill *(below)*, fitted with a carbide-tipped bit. This tool can pound a bit into masonry about 3,000 times per minute. A depth rod indicates when the hole is deep enough.

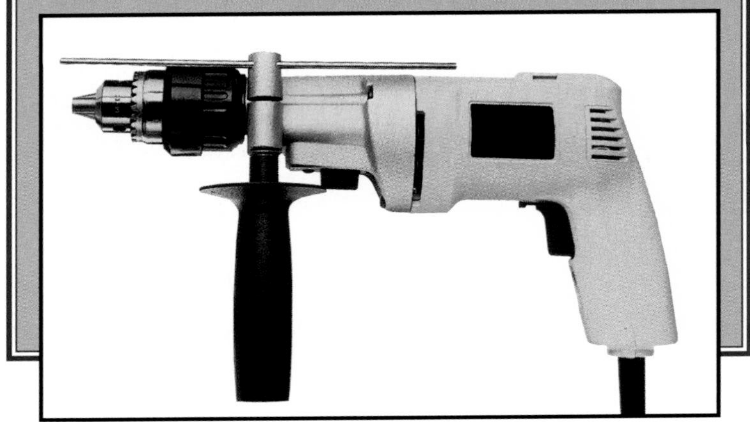

ANCHORING STUDS WITH BOLTS

1. Setting a bolt in epoxy.

◆ Mark and drill holes for the bolts, then dust off the masonry near the holes.
◆ Insert a bolt headfirst into each hole and, with a putty knife, fill the space around the bolt with epoxy.
◆ Cut a strip of 2-inch-wide masking tape for each bolt and, with a utility knife, slice Xs in the center of the strips.
◆ Stick the tape firmly to the wall, with the bolt projecting through the X *(right)*; the tape keeps the bolt centered in the hole and the epoxy from oozing out.
◆ Allow the epoxy to cure for the length of time specified by the manufacturer.

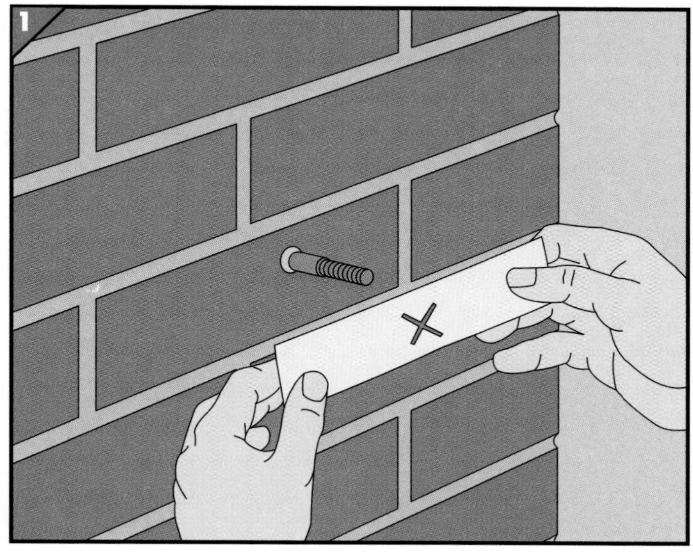

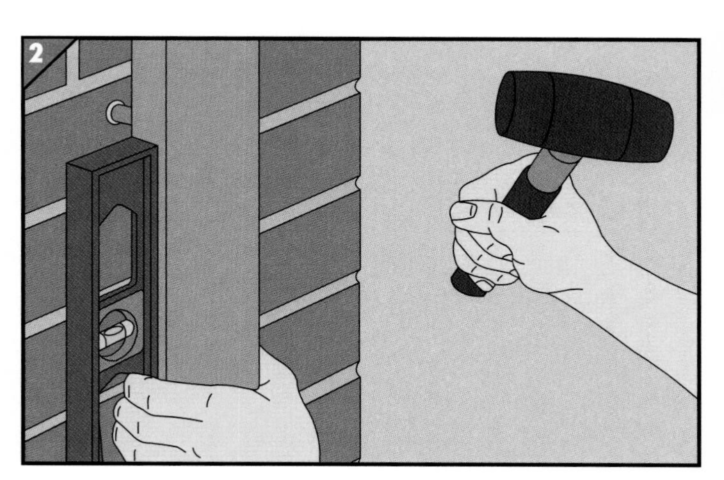

2. Fastening a stud to the bolts.

◆ Holding one side of the board you are mounting against the bolts, plumb the board with a carpenter's level.
◆ Over each bolt, strike the board sharply with a rubber mallet *(left)*, indenting the wood slightly.
◆ Drill a hole through the board at each mark.
◆ Fasten the board to the bolts with washers and nuts.

A POWDER-ACTUATED HAMMER

Loading and firing.

◆ Push the fastener into the muzzle *(right, top)*.
◆ Insert the power load into the chamber.
◆ Hold the tool at 90 degrees to the surface and push the muzzle against the surface, compressing the spring.
◆ Strike the firing pin straight on with a sharp blow from a 1-pound hammer *(right, bottom)*.

CAUTION *This tool contains explosives —keep it out of the reach of children.*

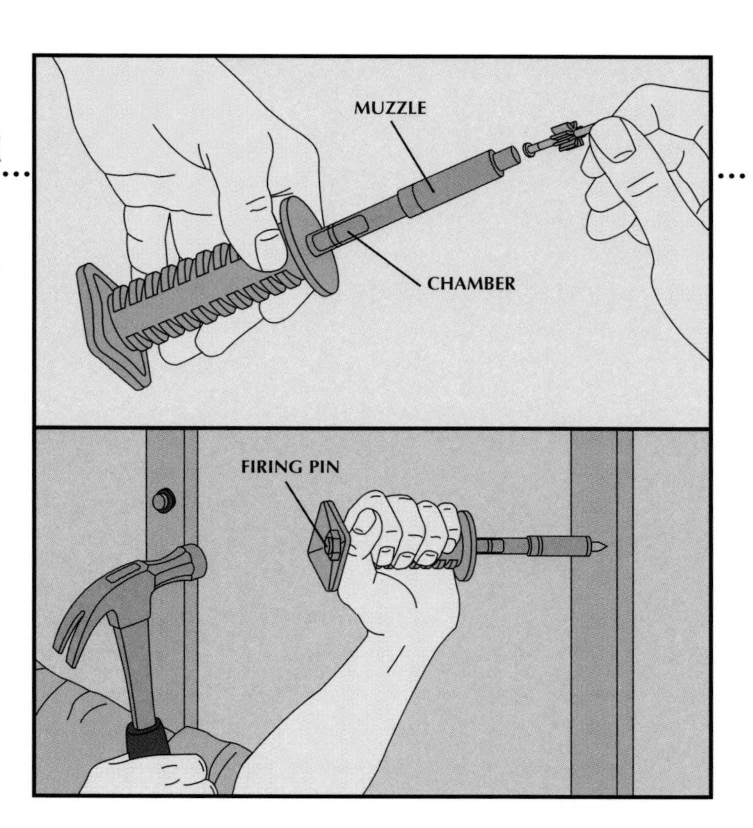

Concrete

Concrete is the most versatile masonry material—it can be molded into any shape within the forms you build. The basic slab-casting techniques found in this chapter can be applied to create a number of projects ranging from a patio to a set of stairs. And the final result need not be drab—a simple slab can be broken up into a grid, colored, or textured in a number of different ways.

Rounding the slab with an edger →

What Goes Wrong and Why

When a concrete slab is well poured and finished, it is nearly indestructible; but incorrect preparation, pouring, or finishing can result in the surface defects shown in these photographs. While most problems can be detected at a glance, dusting is most readily identified by touching the surface with your finger to see if the deteriorated cement is easily picked up.

A common cause of surface defects is an incorrect concrete mixture. Too much water or too much cement will weaken the entire slab. If you use aggregate containing soft stones or clay lumps instead of hard gravel, the surface may break down under normal wear and weather.

Excessive floating or darbying is a primary cause of surface failure. Such overworking sends aggregate toward the bottom of the slab and brings too much water and cement toward the top. When the top of the slab contains too little aggregate, which provides strength, the surface may break up.

Another cause of surface damage is improper curing—letting a slab dry out or freeze and thaw too soon after the concrete has been poured.

Cracking.
Large cracks like those at right open up in concrete that contains too much water, or that was poured too rapidly to be compacted properly. You can avoid cracking by forcing freshly poured concrete into all corners of the form with a shovel. Do not let a batch of concrete dry before pouring the next batch against it—cracks may appear at the junction. Proper compaction of the base prior to the pour and control joints added to the fresh concrete also help prevent cracking. Cracks in concrete can be repaired as described on page 19.

Scaling.
A concrete mixture containing too much water will lack strength once it has cured, causing the top layer to crumble or scale *(left)*. Scaling also occurs where concrete with no air-entrainment additive is used in a freezing climate, or where a slab is subjected to freezing and thawing or deicing compounds before it has cured properly. Deicing products containing ammonium sulfate or ammonium nitrate can sometimes cause scaling even when a slab has cured properly.

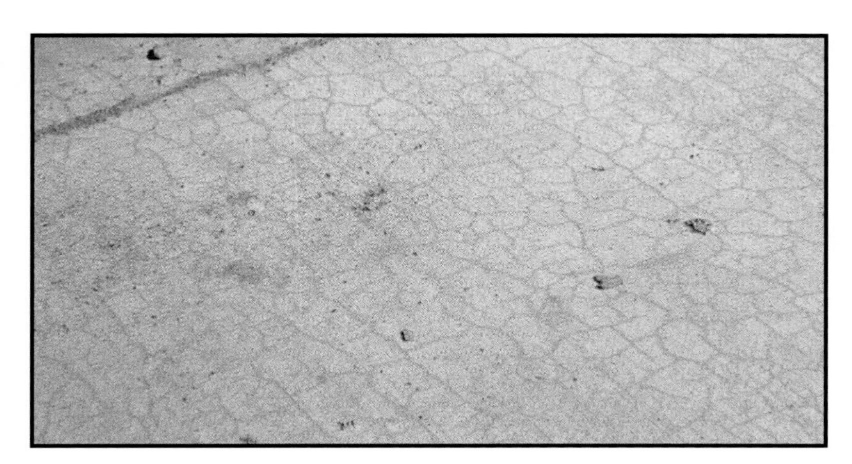

Crazing.
A network of hairline cracks may appear on the surface of a concrete slab that contains too much cement *(left)*; such mixtures shrink excessively as they dry. Improper curing will also cause crazing.

Popouts.
If the aggregate contains lumps of clay or crumbly stones, these soft elements will deteriorate and wash away when the concrete has dried, leaving surface holes called popouts *(right)*. Popouts can be filled with latex or epoxy concrete-patching compounds, or mortar.

Spalling.
If the surface of a slab is weakened by too much darbying or troweling, thin layers of concrete will flake away—a condition called spalling *(left)*. You can patch a spalled surface with an epoxy concrete-patching compound, as illustrated on page 21.

Dusting.
Improper curing or overworking a slab with a darby or float can weaken the top layer of a slab to such an extent that you will be able to pick up powdered cement from the surface on your finger *(right)*. To halt this deterioration, brush the surface thoroughly to remove loose material, then apply a concrete sealer—an acrylic-polymer compound available at most paint stores. One gallon will cover about 400 square feet.

Preparing for the Project

Before starting a job, check your local building codes. Codes and zoning laws may dictate the project's dimensions, location, and design; as well as the quality standards the materials must meet. In some cases, you may need a building permit. Adhering to these rules can often help you avoid costly mistakes.

Inspecting the Site: When you are designing a walk, wall, or patio, take into account the slope of the ground; any rock outcroppings, ponds, or streams; the trees, shrubs, and their roots; and the location of gas, electric, water, or sewage lines and of dry wells, septic tanks, or cesspools—including abandoned ones. Then draw a scaled layout that shows existing structures and landscape features.

Testing the Soil: For a concrete slab, dig a test hole about 1 foot deep and inspect the soil. Unless the earth is very sandy and well drained, the slab will need a 4-inch gravel or sand drainage bed under it to keep the concrete dry and permit it to shift without cracking when the earth freezes and thaws.

Footings for walls *(pages 63-64)* must extend below the frost line. Dig a hole to this depth to check the soil. For both slabs and footings, the ground must be stable. Avoid any site with more than 3 feet of recent landfill. If you strike water in the test hole, or if you live in an earthquake zone, consult a professional.

Preparing the Ground: Before excavating, clear the area by moving plants and, if necessary, getting rid of old concrete. To break up concrete that is not reinforced with wire mesh, lift and drop a heavy sledgehammer onto it, working from the edges toward the center. For a large area, consider renting a jackhammer; call in a professional for reinforced concrete. With the site cleared, consult your preliminary drawings and lay out the design with a string or garden hose.

Ordering Materials: For a small project, or one you plan to divide into sections with permanent forms, you can make your own concrete *(pages 26-28)*. For projects requiring more than about 2 cubic yards of concrete, you may want to order ready-mix concrete *(opposite)*.

Ready-mix concrete, gravel, and sand are sold by the cubic yard. Estimate the length, width, and depth of the project and calculate its volume, keeping in mind that there are 27 cubic feet in a cubic yard. Or, simply divide the cubic feet by 25—resulting in cubic yardage with an allowance of about 8 percent for waste. The chart below provides the number of cubic yards of materials required for both 4- and 6-inch-thick slabs or drainage beds.

ESTIMATING MATERIALS FOR CONCRETE SLABS

Area of slab	Thickness of slab 4 in.	6 in.
10 sq. ft.	.12 cu. yd.	.18 cu. yd.
25 sq. ft.	.32 cu. yd.	.48 cu. yd.
50 sq. ft.	.64 cu. yd.	.96 cu. yd.
100 sq. ft.	1.28 cu. yd.	1.92 cu. yd.
200 sq. ft.	2.56 cu. yd.	3.84 cu. yd.
300 sq. ft.	3.84 cu. yd.	5.76 cu. yd.
400 sq. ft.	5.12 cu. yd.	7.68 cu. yd.
500 sq. ft.	6.4 cu. yd.	9.6 cu. yd.

Calculating cubic yards.
For most residential applications, a 4-inch slab is adequate; a slab that is thinner than this may crack. If you are building a driveway that will be used by delivery trucks, make the slab 6 inches thick. Once you have determined the desired area of the slab, use the table at left to determine the number of cubic yards of concrete, gravel, or sand required for a slab or drainage bed either 4 or 6 inches thick. Add about 8 percent for waste and spillage.

ORDERING READY-MIX CONCRETE

For large jobs, it is easiest to buy the concrete from a ready-mix company that will make it to your specifications and deliver it ready to be poured. But when the truck arrives, you will have to work fast and probably enlist several assistants.

When ordering the concrete, tell the dealer how many cubic yards you need *(opposite)* and describe how you want to use it. In industry jargon, most jobs will call for $5\frac{1}{2}$- to 6-bag concrete—made with about 520 to 560 pounds of cement for each cubic yard. Where winters are severe, use 6-bag concrete; elsewhere $5\frac{1}{2}$-bag is adequate.

Professionals measure the consistency of wet concrete by the number of inches a cone-shaped mass will slump when a conical form is lifted off. You will need concrete with a 4- to 6-inch slump. Concrete with less than 4 inches of slump is too stiff to be workable; with more than 6 inches, it becomes soupy and the concrete will fail. Also, ask for a coarse aggregate (gravel) with a maximum size of 1 inch diameter and for a 6 percent air entrainment *(page 26)* in freezing climates; 4 percent elsewhere.

How much you pay depends on the size of your order. If you need more than 3 to 4 cubic yards of concrete, you may find that ready-mix costs less than mixing the concrete yourself. If you need less than this, you may have to pay a delivery charge or a higher price per cubic yard. Most prices include 30 to 45 minutes' delivery time. If the truck stays longer while you unload it, you will generally pay an hourly rate for the extra time.

To speed up delivery, decide in advance where the truck will stop and how you will get the concrete to the work site. Having the truck drive up onto your property is seldom advisable. The weight of the truck can break a curb, driveway, or sidewalk and make deep ruts in your lawn. If the time and work saved are worth the risk, however, you can minimize the damage by laying 2-inch-thick planks along the truck's route to equalize the load. However, never let the truck drive over a septic tank or cesspool.

Usually, the best plan is to have the truck park on the street and use chutes and wheelbarrows to unload the concrete. A 10- to 12-foot chute is standard equipment on most trucks. Some companies also provide large hoses to pump concrete from the truck directly to the building site. This is an expensive option, but handy for hard-to-reach sites.

CAUTION *A chute or wheelbarrow full of concrete is extremely heavy. Agree on signals with the truck operator and your helpers before the pour begins, and keep children away from the truck.*

Unloading ready-mix.

◆ Organize a crew of helpers ahead of time. Have some of them place the fresh concrete, and others smooth and finish the slab.

◆ Set a 4-by-8 plywood panel where the truck will discharge the concrete. Then place 2-foot-wide plywood strips or 1-by-8 planks from the unloading area to your work site to make a track for the wheelbarrows—the type with a pneumatic tire—and prevent the wheels from sinking into the lawn.

◆ When the truck arrives, work quickly with a shovel to slide the concrete from the delivery chute into each wheelbarrow *(above)*, taking care not to overfill it. Do not leave concrete in the chute—it may drop out before you return. As you work, shift the wood track to the most convenient point for dumping the wheelbarrow loads.

Forms for Slabs

Forms serve as molds for concrete, holding it to shape as it cures. Generally, forms are assembled on site before the concrete is poured and removed after it has cured; but some can remain in the slab as permanent decorative features *(page 59)*.

Staking and Grading: The first step is to lay out the slab with stakes. These are cut 18 inches long from 1-by-2s or 2-by-2s. The stakes support the forms and indicate the grade, or pitch, of the slab, which ensures it will shed water. For a walkway that will run in a straight line over ground that slopes gently away from the house, follow the natural slope. If the ground doesn't slope at all, grade the slab from side to side to ensure proper drainage. A lot that slopes abruptly requires a slab with steps.

Excavating: With the stakes in place, the site is excavated, leveled, and compacted. The required depth of the excavation depends on local codes; but for structures like walkways, patios, steps, and driveways, 8 inches is generally sufficient to accommodate the thickness of the slab and 4 inches of gravel beneath it.

Building Forms: A straight form for a simple 4-inch-thick slab, such as a walkway *(below)* or patio, is made from 1-by-4s or 2-by-4s. For curved sections, flashing metal can be used *(page 45)*.

The best choices for temporary form boards are fir, pine, or spruce. Buy wood that is planed smooth on all its surfaces and, unlike lumber used in most construction, green (unseasoned). Fully dried wood may absorb moisture from the concrete and interfere with curing. Permanent forms can be made of redwood, cedar, or pressure-treated wood. Use single boards for each side of the form, if possible, but long sections can be pieced together by butting boards end to end and reinforcing the seams with stakes.

Gravel and Expansion Joints: After the forms are assembled, a gravel drainage bed is laid in place and leveled *(page 46)*. Before the concrete is poured, expansion-joint material is added against any adjoining structure, such as steps, a wall, or another slab, so the new slab can shift independently *(page 46)*.

TOOLS

Ruler
Maul
Shovel
Flat spade
Rake
Tamper
Line level
Mason's level
Hammer
Handsaw
Circular saw
Tin snips

MATERIALS

1 x 4s, 1 x 8
2 x 4s
Stakes
String
Sand
Gravel
Common nails ($\frac{1}{2}$", 3")
Flashing metal
Expansion-joint material

SAFETY TIPS

Protect your eyes with goggles when hammering or driving in stakes.

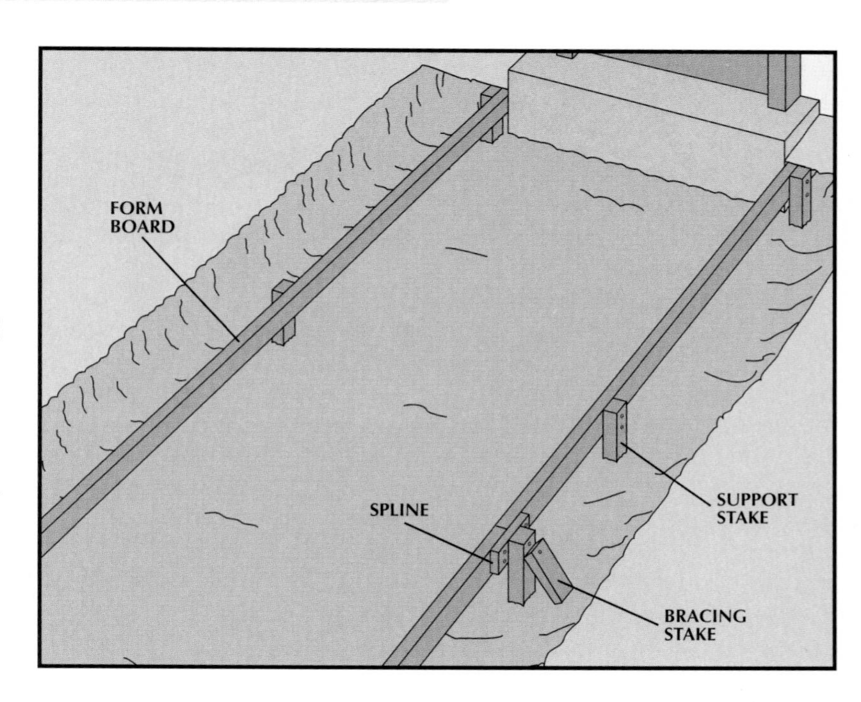

Anatomy of a simple form.
After the walk is outlined and the site excavated, support stakes are set at both ends—$\frac{3}{4}$ inch or $1\frac{1}{2}$ inches outside the proposed slab to allow space for 1-by-4 or 2-by-4 form boards—and at 3- to 4-foot intervals in between. Form boards are nailed to the stakes, leaving room for a 4-inch gravel bed under the slab. The stakes are sawed off even with the top of the form boards. The joints between form boards are reinforced with splines and with diagonally set bracing stakes.

EXCAVATING THE BASE

1. Outlining the slab.

◆ For a straight slab, drive stakes at each corner of the proposed slab with a maul.
◆ Tie a string to each stake at one end of the proposed slab and run it to the stake at the opposite end.
◆ Mark the string lines on the ground by trickling an unbroken line of sand along each one *(left)*.

For a curved slab, use the technique on page 45 to position stakes around the curve.

2. Digging the trench.

◆ Remove the stakes and strings.
◆ Dig a trench of the required depth to about 1 foot outside the sand lines *(right)*. If you plan to reuse the turf later along the edge of the finished slab, cut it out with a flat spade and save it.

3. Leveling the base.

◆ Smooth the base of the trench with a rake.

◆ Remove any large rocks and fill in the holes with sand or gravel.

◆ Flatten the base by pulling a 2-by-4 as long as the base is wide across the surface *(left)*.

4. Tamping the base.

◆ Compact the soil with a cast iron tamper *(right)*. For a large project, rent a vibrating tamper *(page 56)*.

◆ Fill low spots with gravel and tamp again.

◆ With the 2-by-4, check that the bottom of the trench is even across its width; add gravel and tamp again as necessary.

STAKING THE SITE

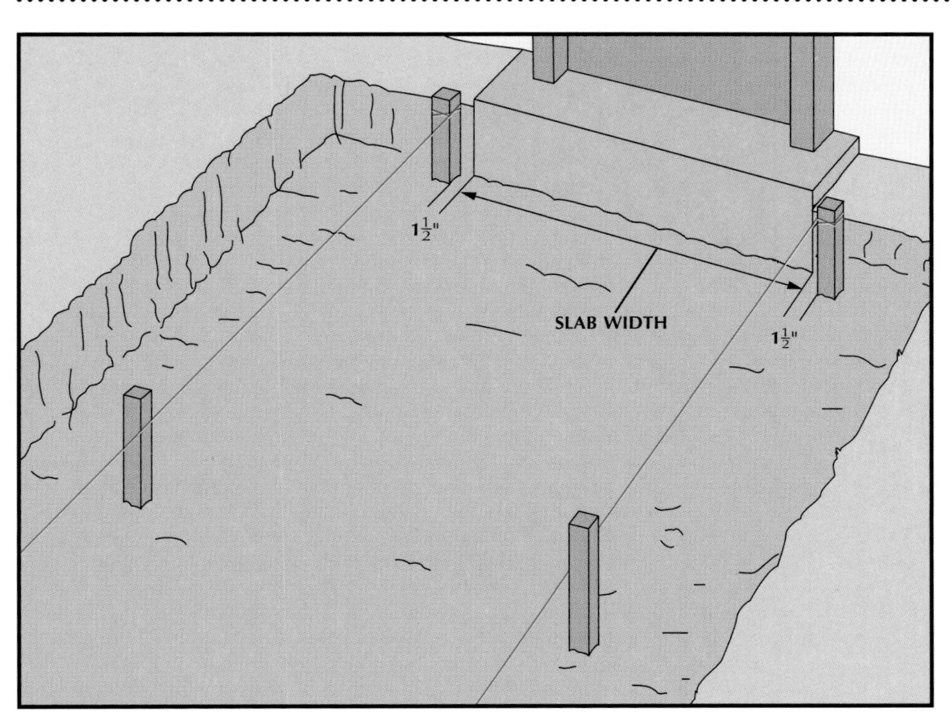

Driving support stakes.

◆ Drive the corner stakes again—this time about 6 inches deep—positioning them outside the desired slab by the thickness of the form boards (1-by-4s or 2-by-4s).

◆ Mark the end stakes at each end of the slab at ground level. Tie strings between the end stakes, running the strings along the insides of the stakes.

◆ Drive support stakes at 3- to 4-foot intervals between the end stakes along both sides of the slab so their insides lie against the strings, as shown at left. If there is a curve in the walkway, do not set the stakes for the curve until the form boards for the straight sections have been added.

GRADING A WALKWAY ON A SLOPING LOT

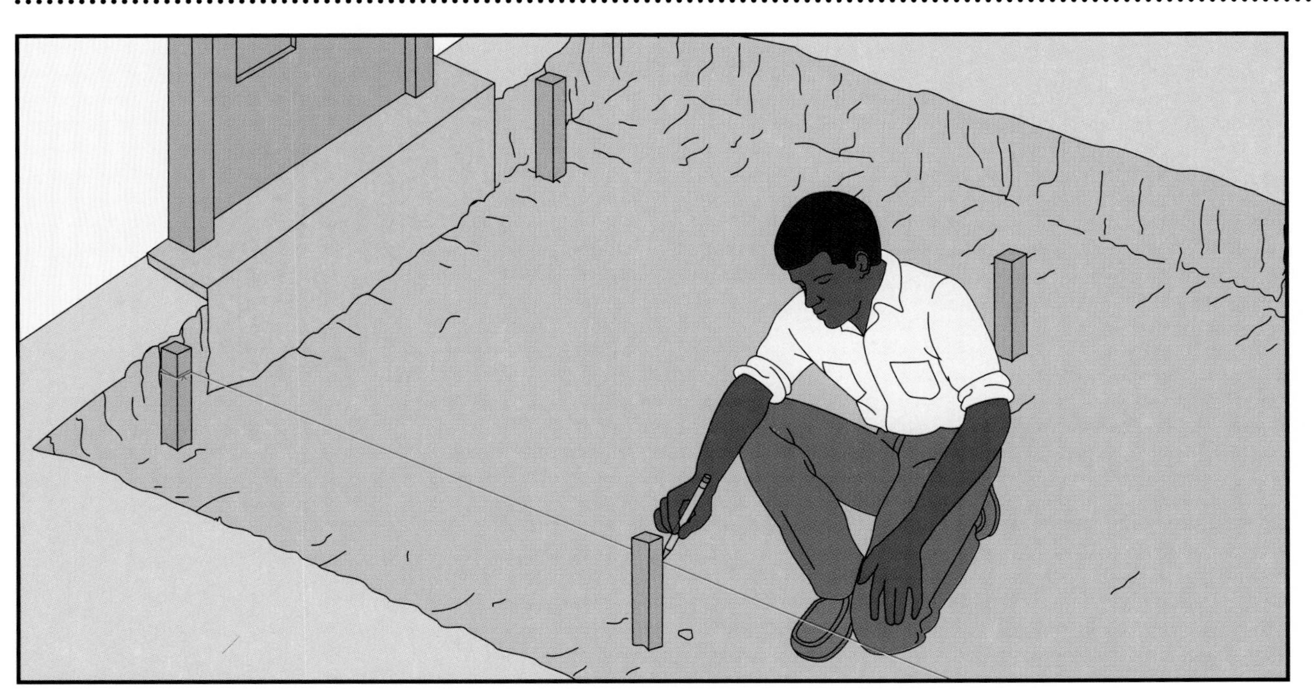

Marking the lengthwise grade.
The form boards on a gently sloping site can follow the slope of the land. However, you'll need to compensate for minor rises and dips:

◆ Mark the intermediary stakes level with the string lines *(above)*.

◆ Repeat for the stakes along the other side of the walkway.

SLOPING A WALKWAY ON A FLAT SITE

1. Leveling one set of stakes.
◆ Hang a line level at the middle of the string and raise the low end of the string until the string is level, and at ground level.
◆ Mark the intermediary stakes level with the string.

2. Grading the slab across its width.
◆ Nail one end of a long straight board the width of the slab to the marked end stake, level with the mark. Use a single nail so the board can pivot.
◆ Place a level on the board to level it and mark the opposite stake *(above)*.
◆ Remove the board and mark a second line below the one you just marked so the slab will slope from side to side $\frac{1}{4}$ inch for each foot of width.
◆ Repeat for the pair of stakes at the opposite end of the walkway.
◆ Run a string between the two end stakes you've just marked and mark the intermediary stakes level with the string.

ATTACHING FORM BOARDS

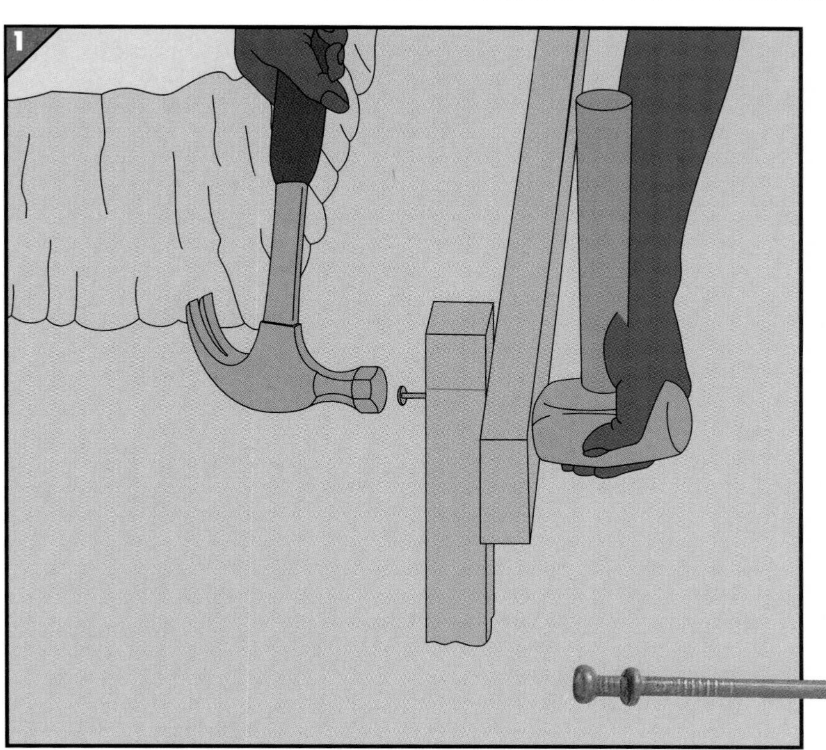

1. Nailing the boards to the stakes.

◆ Starting on one side of the excavation, hold an end of a form board against the inside edge of the stake nearest the house—flush with the top of the stake on sloping ground; flush with the grading line on flat ground.
◆ With a 3-inch common nail—not galvanized, so it can be pulled easily—fasten the board to the stake, cushioning the blows with a maul *(left)*. Alternatively, you can use double-headed nails *(photograph)*, which are easily removed.
◆ Nail the board to the stake at its opposite end, then to the stake at its middle.
◆ Drive a second set of nails below the first.
◆ Attach the board to any intermediate stakes.
◆ Where form boards meet end to end, nail a spline across the joint on the outside faces of the boards, then drive an additional support stake behind the spline as shown in the illustration on page 38.

2. Making corners.

◆ At a corner, butt the form board against the back face of the one already in place.
◆ Fasten the board to the support stake, holding its top edge level with the first board *(right)*.

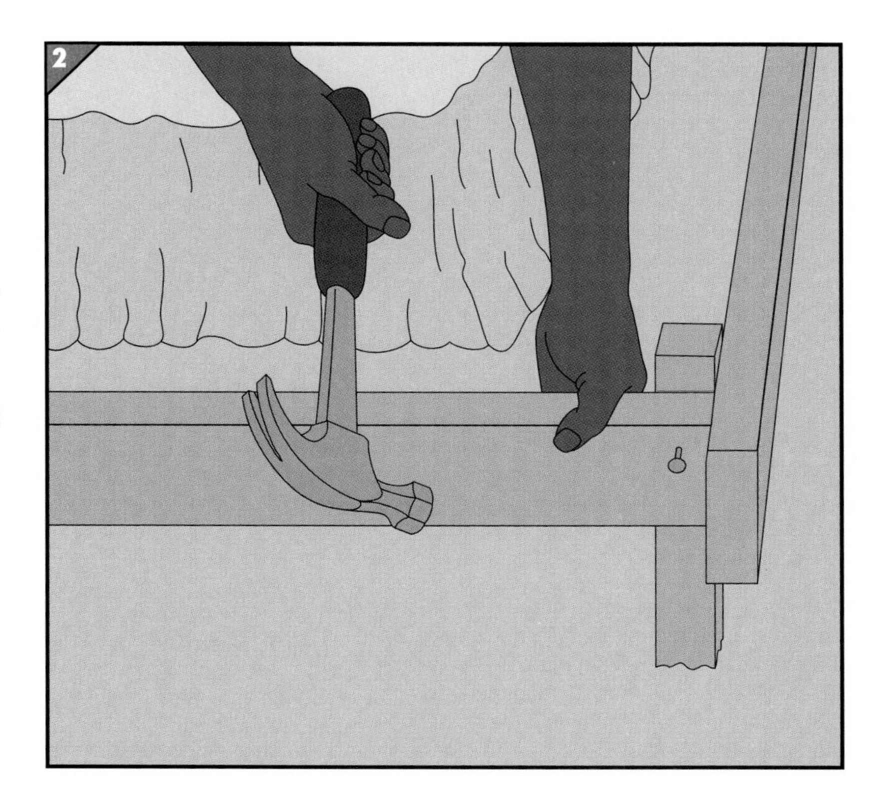

3. Adding braces.

◆ At each corner and end joint between form boards, drive a bracing stake into the ground on an angle so it presses against the back of the support stake below the nails.

◆ Nail the braces to the support stakes *(right)*.

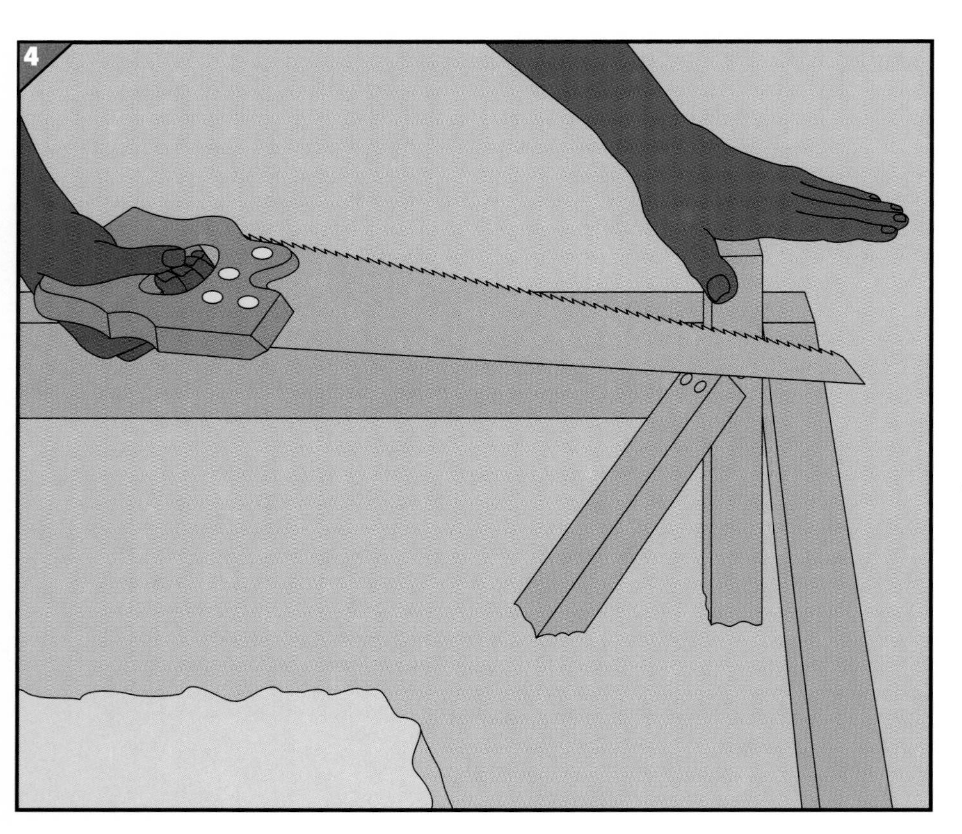

4. Trimming the stakes.

Cut off the part of the support stakes that extends above form boards to provide a uniformly flat surface for screeding the slab later.

SHAPING CURVES WITH METAL

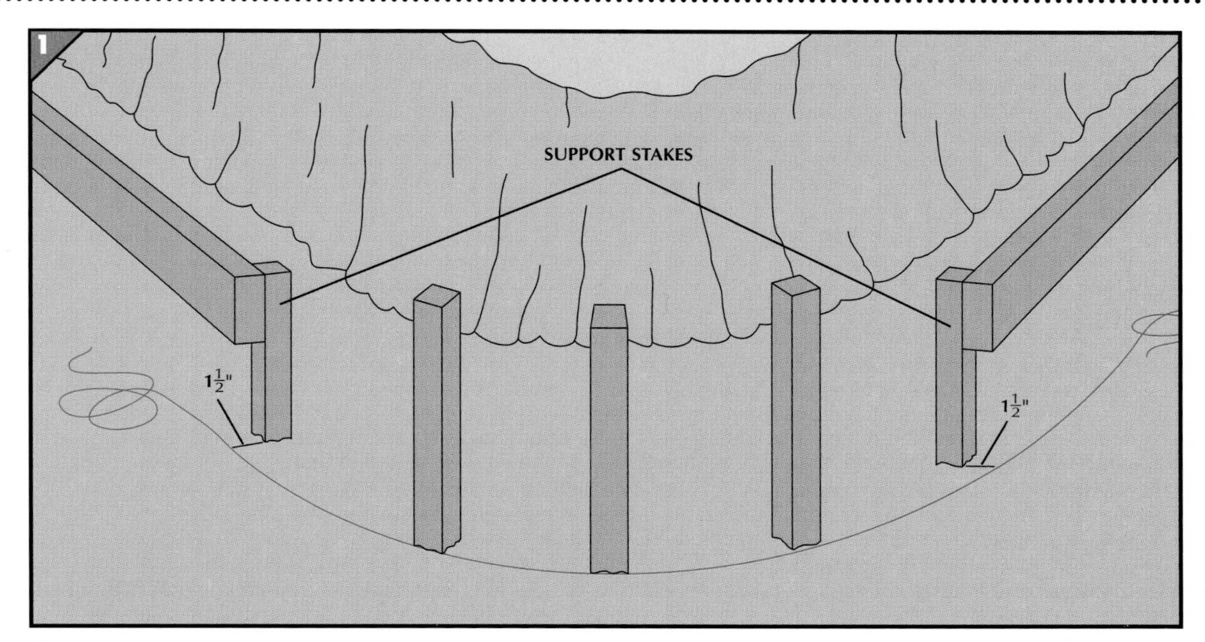

1. Staking curves.

◆ Lay string or a garden hose in the desired arc $1\frac{1}{2}$ inches inside the support stakes at the start and end of the curve.

◆ Drive support stakes for the curve about every foot along the string or hose, as shown above.

◆ Tie a string between the stakes at each end of the curve so it is level with the grading marks on the stakes.

◆ Pull the string taut and mark a grading line on each stake along the curve where the string crosses it, then remove the string.

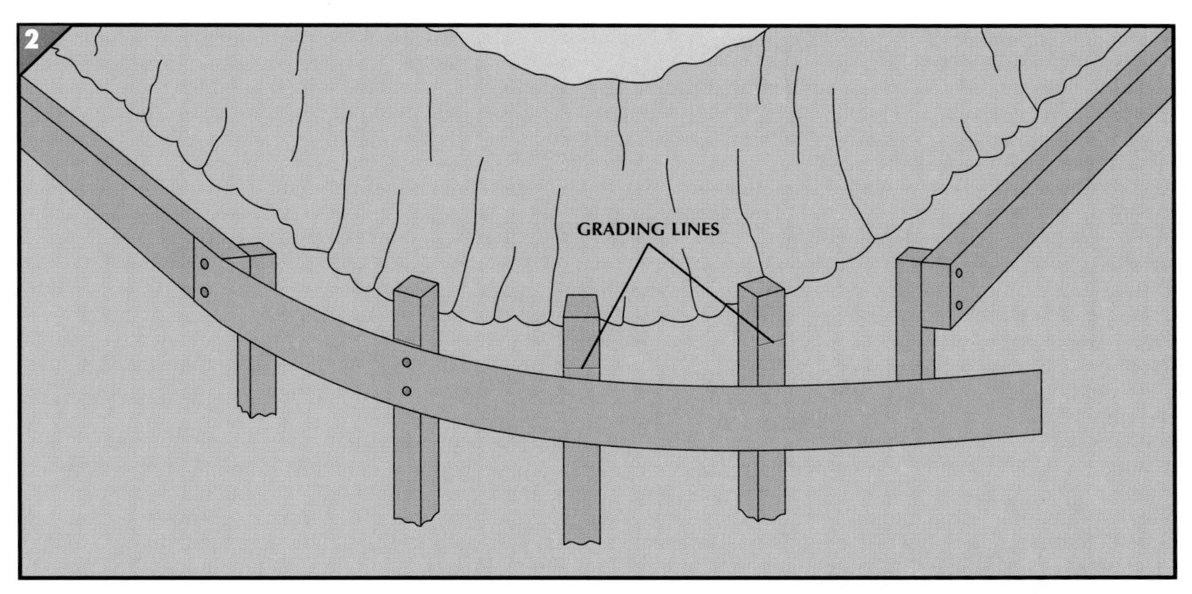

2. Attaching curved forms.

◆ With tin snips, cut a strip of flashing metal 4 inches wide and 6 inches longer than the curve.

◆ Overlapping the form boards at the ends of the curve by 3 inches, line up the top of the strip of metal with the boards and the grading marks on the support stakes. Attach the strip to the stakes with $\frac{1}{2}$-inch common nails.

◆ Add a bracing stake behind each support stake *(page 44, Step 3)*, then trim the tops of the support stakes level with the form boards and strip of metal.

◆ Backfill the trench outside the curve for extra support, taking care not to tamp the earth so hard that you move the stakes.

LAYING A DRAINAGE BED

Placing gravel.

◆ Shovel a 4-inch layer of gravel into the forms.

◆ Make a bladed screed from a 2-by-4 long enough to extend over the form boards and a 1-by-8 or 2-by-8 as long as the space between the form boards. Fasten the boards together as shown above so that the 4-inch "blade" fits down between the form boards.

◆ Level the gravel bed from one end to the other by pulling the screed, blade-down, toward you and running the ends on the form boards *(above)*—the blade will reach down $\frac{1}{2}$ inch lower than the bottom of the form boards. Fill in any low spots with gravel.

◆ Compact the bed with a tamper, adding more gravel if necessary. Then level the surface again with the screed.

DEALING WITH JOINTS

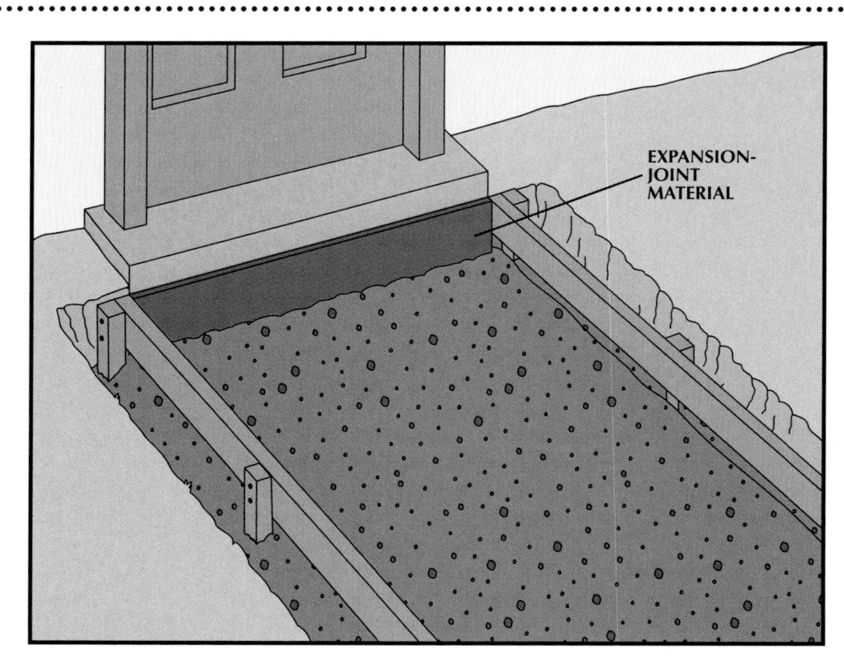

Installing joint filler.

◆ For each structure adjoining the new slab, cut a piece of expansion-joint material—usually asphalt-impregnated fiberboard—to a width of 4 inches and long enough to fit snugly between the form boards.

◆ Place a piece against each structure flush with the top of the form boards, as shown at left.

Pouring and Finishing the Slab

On the day of the pour, have everything ready to go and in place before the ready-mix truck arrives. A fresh mix needs to be poured and finished in about three hours; after that, it becomes too stiff to be workable. In that time, one person, working alone with ready-mix concrete, can place and finish about 30 feet of a 3-foot-wide walkway. One helper will make it possible to lay 50 feet of walk in the same time; two helpers, 70 feet.

Pouring the Concrete: Moisten the forms, and in hot weather, the gravel base as well. Once the concrete is placed, compact, level, and then smooth it. Working the concrete, as described below, forces "bleed" water to the surface; do not work the mixture further until the water disappears. For air-entrained concrete on a hot dry day, the water will likely evaporate completely in 10 to 20 minutes; in cool or humid weather, an hour or more may be needed.

Finishing: Once the bleed water disappears, the surface can be finished and the edges rounded to reduce the chance of chipping *(page 49)*. Incise control joints in the slab *(page 50, Step 2)* at intervals of 24 times the thickness of the slab (every 8 feet for a 4-inch slab) to induce cracks at the joints rather than at random. These joints are not needed in footings. If you want the marks of the edger and jointing tool to be visible, finish the surface with the desired texture first. Otherwise, do the edges and joints before finishing.

Next, finish the slab with a float and trowel *(pages 50-51)*. For a nonslip surface, required for an exterior slab, you can either stop after floating—resulting in a less finished appearance—or trowel the slab and then roughen the surface with a broom *(page 51)*. Other decorative finishes—embedded stones and imitation flagstone—are shown on page 52.

Curing: Keep concrete damp and undisturbed for at least a week after finishing to allow the chemical reactions that give it strength to proceed *(page 53)*. After curing, the forms can be removed, but keep heavy loads off the surface for an additional week. If you plan to paint or stain the concrete, allow it to cure for 6 weeks first.

TOOLS

Square shovel
Flat spade
2 x 4 screed
Darby
Pointing trowel
Edger
Jointer
2 x 8 guide board
Carpenter's square
Magnesium or wood float
Steel trowel
Stiff-bristle broom

MATERIALS

Polyethylene sheeting
Bricks
Burlap sacks

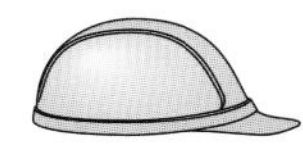

SAFETY TIPS

If you are mixing concrete, put on goggles and a dust mask. Wear goggles when pouring or screeding concrete to protect your eyes from splashes. Wet concrete is caustic—don gloves when working with it. Goggles and rubber gloves protect you from splashes when washing a slab with muriatic acid.

PLACING THE CONCRETE

1. Filling the forms.

◆ Dump the first wheelbarrow load of concrete into the forms—those farthest from the truck if you are using ready-mix. With square shovels, pack the concrete into the corners of the forms, being careful not to knock down or completely submerge the expansion-joint material.
◆ Shovel each successive load up against the preceding one, overfilling the forms by about $\frac{1}{2}$ inch.
◆ Thrust the shovel edge through the concrete as you work it to eliminate air pockets.

2. Clearing the edges.
As soon as one section of forms is completely filled, drive a flat spade down between the concrete and the inside surface of the forms to force large pieces of aggregate away from the edges of the slab *(left)*.

3. Compacting and leveling.
Use the edge of a straight 2-by-4 or 2-by-6, about 2 feet wider than the forms, as a screed. Zigzag the screed from side to side across the slab *(right)*.

If you plan to embed stones in the surface of the slab, follow the instructions on page 52 after screeding.

4. Smoothing the surface.

◆ Working quickly, smooth the concrete and eliminate any "hills" or depressions with a darby. Pressing down lightly on the trailing edge of the darby, sweep it back and forth across the surface in wide arcs to force large aggregate down into the concrete *(left)*.

◆ When bleed water floats to the surface, stop darbying. Wait until the water evaporates and the shiny surface dulls, then immediately begin edging and jointing the slab and finishing the surface *(below)*.

If you plan to create a flagstone finish, do so now, following the instructions on page 52.

FINISHING TOUCHES

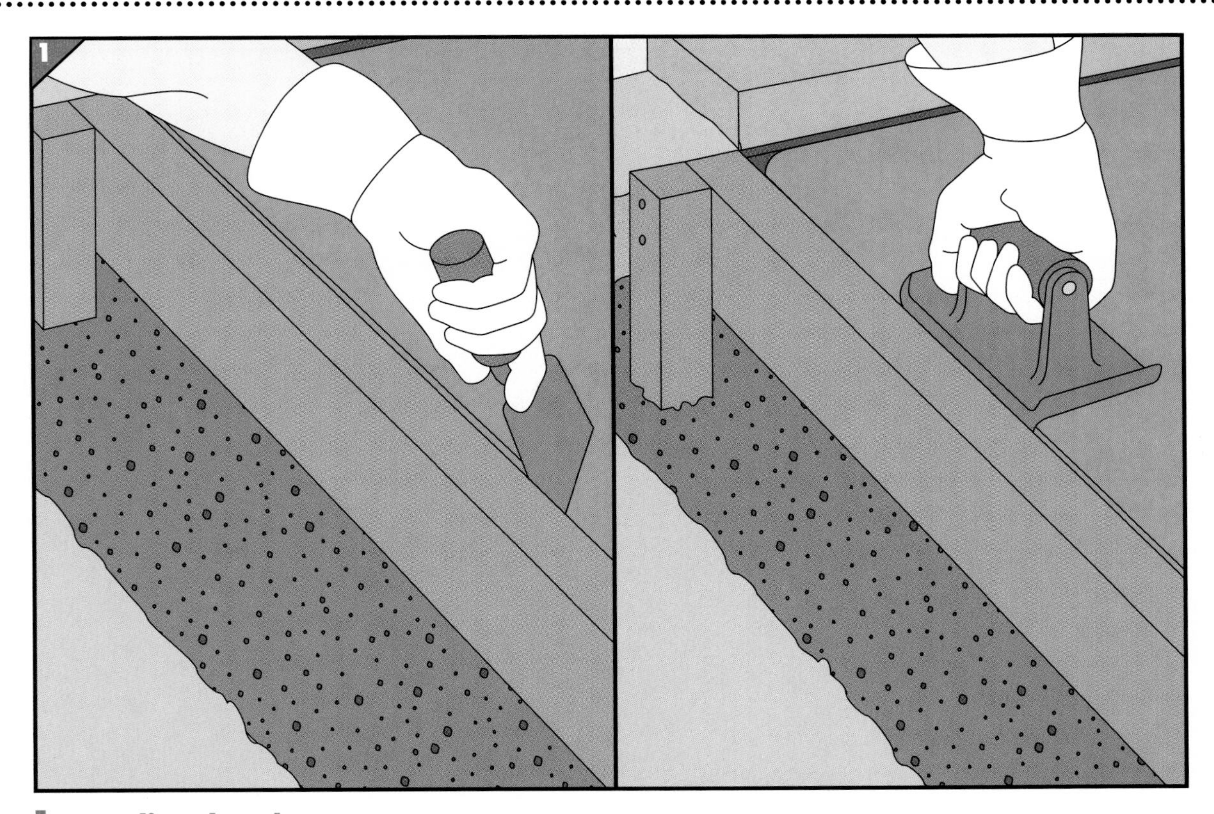

1. Rounding the edges.

◆ Draw a pointing trowel along the inside edges of the forms to cut the top inch or so of concrete away from the wood *(above, left)*.

◆ Finish the sides of the slab by running an edger with a $\frac{1}{2}$-inch radius firmly back and forth along the edges until they are smooth and rounded *(above, right)*.

2. Cutting control joints.

◆ At each point where you wish to produce a joint, set a 2-by-8 as a guide board across the forms; check with a carpenter's square to ensure the board is perpendicular to the edges of the slab.

◆ With a jointer that has a radius of $\frac{1}{4}$ to $\frac{1}{2}$ inch and will cut to a depth of one quarter the slab's thickness—1 inch in this case—press down into the concrete along the edge of the guide board. Run the tool back and forth to cut the joint *(left)*.

3. Floating the surface.

◆ For concrete with an air-entrainment additive, smooth and compact the concrete with a magnesium float; otherwise, use a wood float. Starting at one end of the slab, press the float flat on the surface and sweep the blade back and forth in gentle curves *(right)*. Support yourself on a second float to lean over the concrete, or use kneeboards *(page 59)*.

◆ Move backward as you work toward the slab's opposite end to remove any hand or knee prints in the concrete. Do not overwork the concrete; this will drive aggregate down and bring water and cement to the surface, weakening the slab.

4. Troweling the surface.
To further smooth and compact the surface, use a steel trowel. Keep the blade nearly flat and sweep the trowel back and forth in arcs 2 to 3 feet wide *(right)*.

5. Creating a nonslip surface.
Beginning at one end of the slab, draw a stiff-bristle broom straight across the surface *(left)*; for a curved slab, move the broom in arcs. If the broom picks up small lumps of concrete, hose the broom clean and let the slab dry a few minutes longer before proceeding. If you have to press down heavily to score the surface, work quickly, before the concrete becomes too hard.

DECORATIVE FINISHES

A broomed surface *(page 51)* is only one way to finish a concrete patio or walkway. Shown on this page are two attractive alternatives: embedded multicolored stones and simulated flagstone; the stones help to make the surface more slip-resistant. You can also pattern a concrete surface with commercial stamps that make impressions in the concrete, imitating the outlines of brick, stone, or tile.

Embedding stones.
This finish is easy to achieve by adding stones to a freshly poured concrete slab. Buy round multicolored stones, $\frac{3}{4}$ to $1\frac{1}{4}$ inch in diameter, called aggregate, or $\frac{3}{8}$-inch pea gravel, from a masonry supplier.

After compacting and leveling the concrete *(page 48, Step 3)*, wet the stones and drop them onto the surface at random or in a pattern. Tap the stones down with a darby until the tops are just below the surface *(right)*; if they sink too far, wait for the concrete to firm up. Then, after the bleed water disappears, place a board over the slab. When you can stand on the board without pushing the stones farther down, brush the concrete with a stiff-bristled broom to expose just the top third of each stone. Flush away excess concrete with a fine spray from a hose.

Begin curing the concrete *(page 53)*; after a day or two, uncover the slab and wash the stones with 1 part commercial-strength muriatic acid added to 10 parts water. Continue curing it for another five or six days.

CAUTION *Pour acid into water—never pour water into acid.*

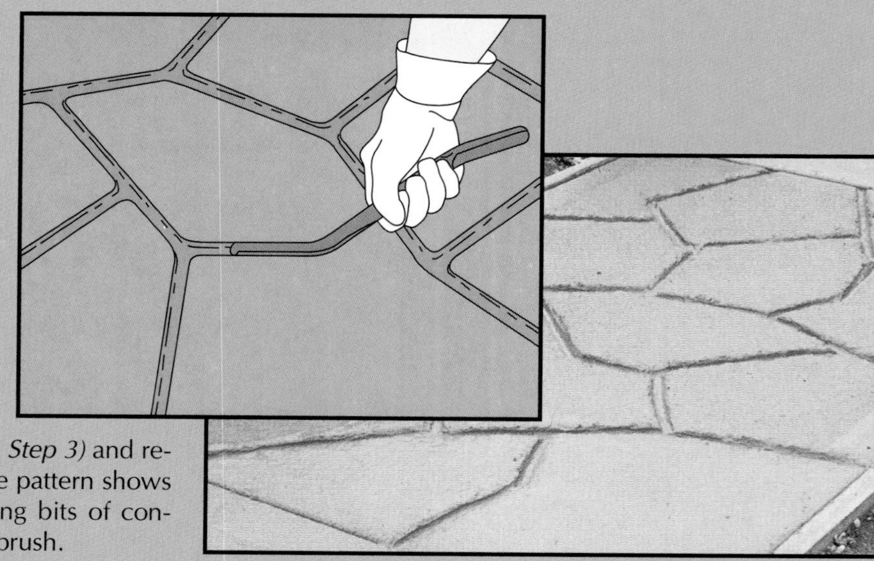

Imitating flagstones.
After smoothing the concrete with a darby *(page 49, Step 4)*, carve irregularly spaced grooves $\frac{1}{2}$ to $\frac{3}{4}$ inch deep into the surface with a convex jointer *(right)*. Alternatively, use a bent length of copper pipe. When the bleed water evaporates, float the concrete *(page 50, Step 3)* and reshape the grooves until the flagstonelike pattern shows distinctly. Then, brush out any remaining bits of concrete from the grooves with a dry paintbrush.

CURING THE SLAB

Keeping the concrete damp.
To ensure proper curing of the concrete, keep it damp for a week. For footings, this can be accomplished by covering the surface with polyethylene sheeting weighted down with bricks; however, this method may stain the surface. For a walk or patio, cover the concrete with clean, soaking-wet burlap sacks and keep the burlap damp by spraying it periodically with a hose *(left)*. Alternatively, use curing compounds, described below.

Once the concrete has cured, disassemble the forms.

SPECIAL PRODUCTS FOR EFFECTIVE CURING

Curing compounds, generally available from a ready-mix supplier, offer an easy and inexpensive way to ensure proper curing of a concrete slab. White wax-based compounds are visible as you apply them, but may stain the surface; resin-based compounds are less visible, but the dyes fade more completely. Some of these products include a sealer; but if you choose one that does not, wait several months after the slab has cured before applying a sealer, paint, or stain.

Apply the compound when the surface is damp, but free of standing water. To cover a large area, use a portable garden-type sprayer; for small projects, a paint roller works well. Cover the surface completely and apply a second coat at a right angle to the first.

Building a Patio

Patios and play areas are built with the same techniques used for small concrete slabs, such as walkways; but they require a larger quantity of concrete. You can place the concrete all at once or in sections. All the information you need to build forms and pour and finish concrete, as well as the tools and materials required, is found on pages 36 to 53.

Layout and Grading: The key to an attractive patio is proper layout. The triangulation method *(opposite)* will ensure square corners. As with a walkway, a patio must slope away from the house *(page 56, Step 1)*.

Temporary Forms: To complete a patio in one day, have the concrete delivered by a ready-mix truck *(page 37)*. Even if you are pouring the concrete all at once, divide a large patio into rectangles with forms *(page 57, Step 2)* so you can level, compact, and screed one section at a time. The center board is then removed and the gap filled.

Permanent Forms: If you prefer to mix your own concrete and pour it in small batches, you can construct a patio with permanent forms that are filled one at a time *(pages 59-60)*. Use decay-resistant wood—cedar, redwood, or pressure-treated lumber—for the forms.

Reinforcement: Not all slabs require reinforcement, but if you plan to surface the slab with brick, tile, or stone, lay wire mesh in the forms before you pour the concrete *(page 57, Step 3)*.

TOOLS

- Tape measure
- Mason's level
- Maul
- Screwdriver
- Hammer
- Shovel
- Tamper
- Lumber for bladed screed
- 1 x 12 and 2 x 2 for bull float

MATERIALS

- Stakes
- String
- Expansion-joint material
- Gravel
- Wood screws (No. 8)
- Wire mesh
- Pressure-treated 2 x 4s
- Galvanized common nails (3")
- Heavy-duty masking tape

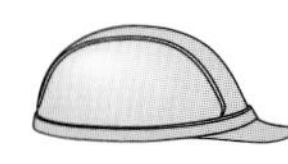

SAFETY TIPS

Put on goggles and a dust mask when mixing concrete. Wear gloves, rubber boots, and goggles when pouring or spreading concrete. Add a hard hat while unrolling wire-mesh reinforcement.

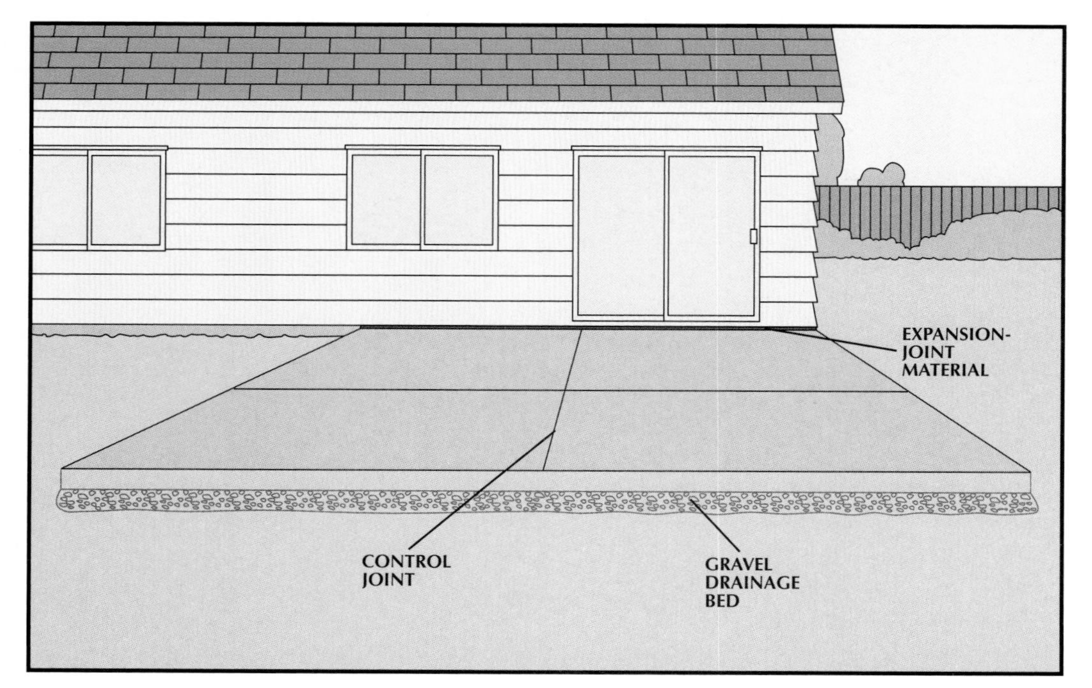

Anatomy of a slab. A typical patio has a 4-inch gravel bed topped by 4 inches of concrete. Control joints divide the slab into a maximum of 8-foot squares, and expansion-joint material is placed between the slab and the house. The patio is sloped so that water will drain away from the house.

LAYING OUT THE DESIGN

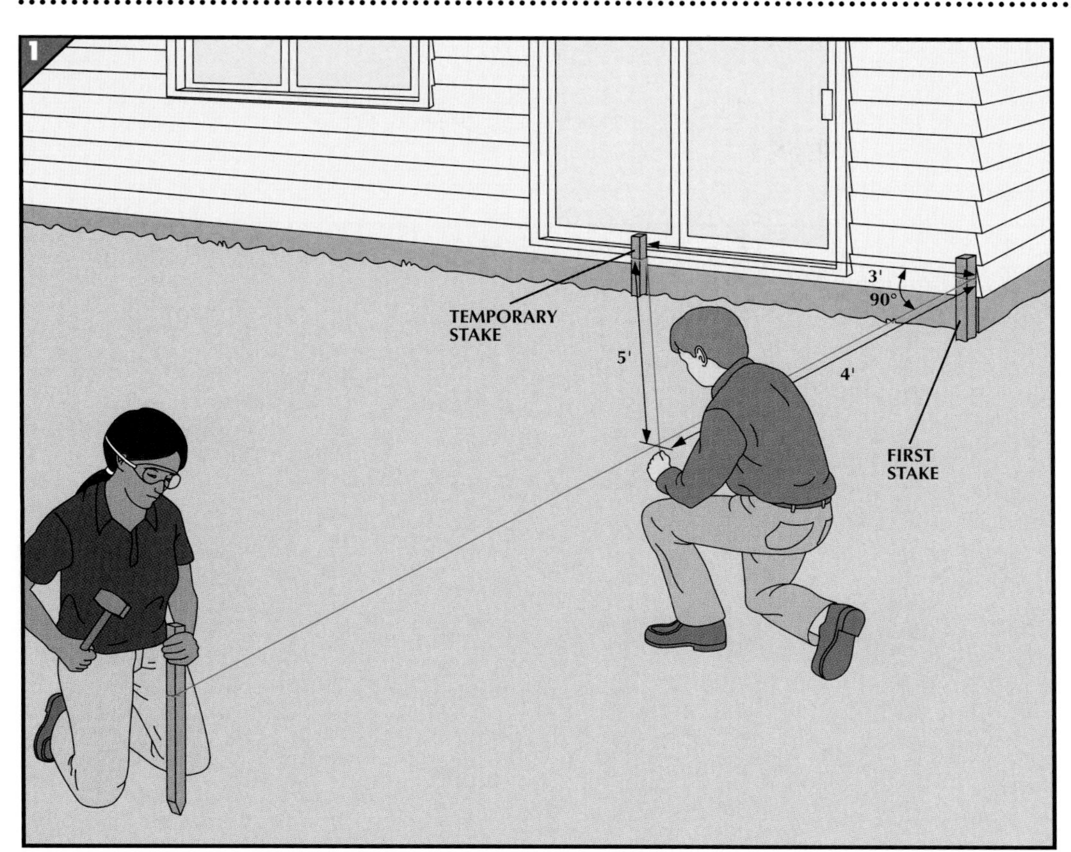

1. Marking the sides.

◆ Drive a stake next to the house to locate one corner of the patio.

◆ Tie a long string to the stake, mark the string at 4 feet, and tie another stake to the string about 1 foot beyond the planned edge of the patio.

◆ Drive a temporary stake along the house 3 feet from the first one and attach a string to it. Mark this string at 5 feet.

◆ Pull the strings together so the marks meet between the stakes; drive the stake tied to the first string into the ground *(left)*, marking one side of the patio.

◆ Remove the temporary stake, locate the second corner of the patio next to the house, and repeat to mark the opposite side.

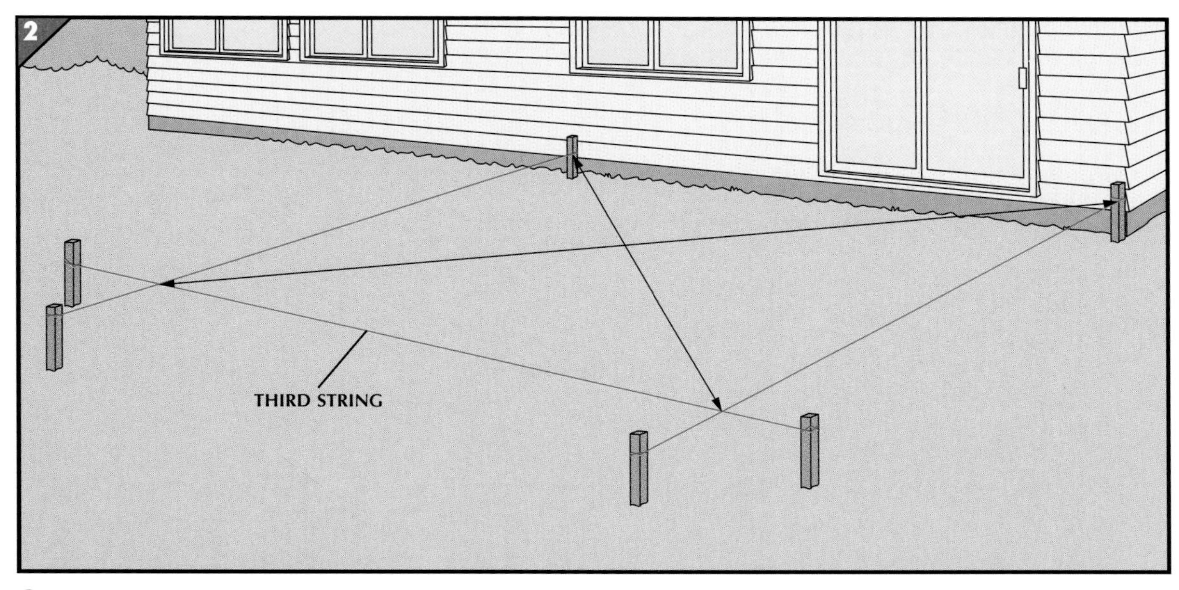

2. Completing the layout.

◆ Mark the desired length of the patio sides on both strings.

◆ Tie a stake to each end of a third string and drive the stakes so the string crosses both length marks.

◆ Check that the layout is square by measuring the diagonals between opposite corners *(black arrows)*—they should be equal. If not, adjust the location of the stakes attached to the third string and recheck.

BUILDING THE PATIO

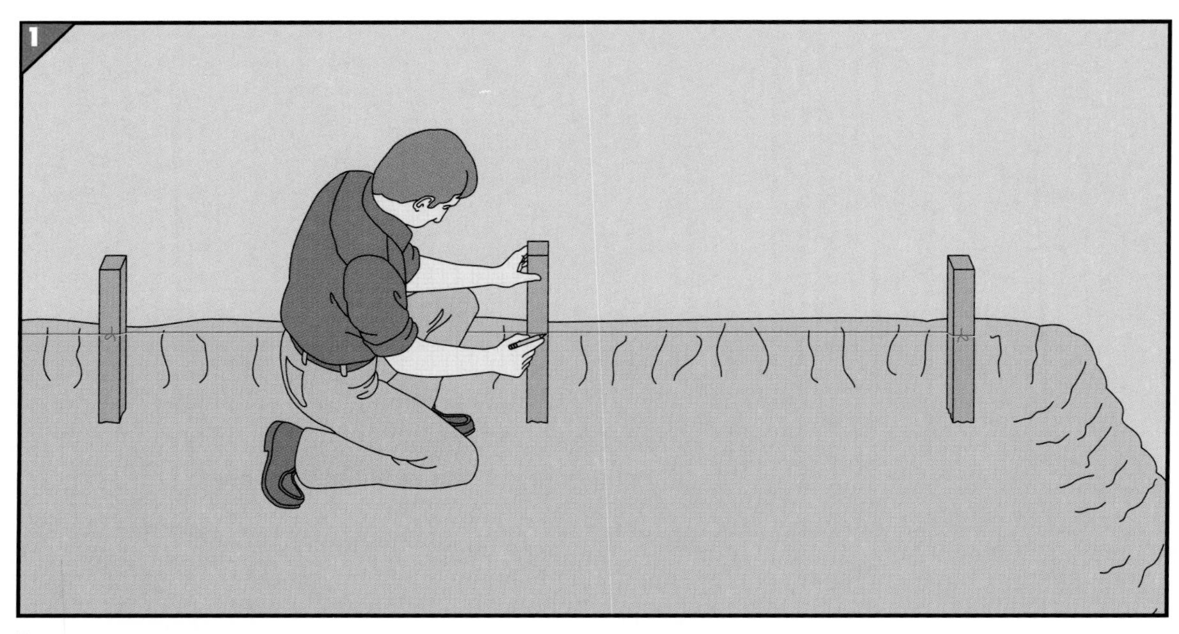

1. Grading the patio.

◆ Following the instructions for a walkway *(pages 39-41)*, transfer the string lines to the ground with sand, excavate the site, and drive support stakes every 3 to 4 feet on three sides of the layout, but not along the house.

◆ Mark the corner stakes at ground level, and run strings between the corner stakes, lining up the strings with the marks. Mark the intermediate stakes level with the string.

◆ Attach form boards to the stakes *(pages 43-45)*, aligning the tops of the boards with the grading marks.

◆ Level and tamp the soil with a vibratory tamper *(below)*.

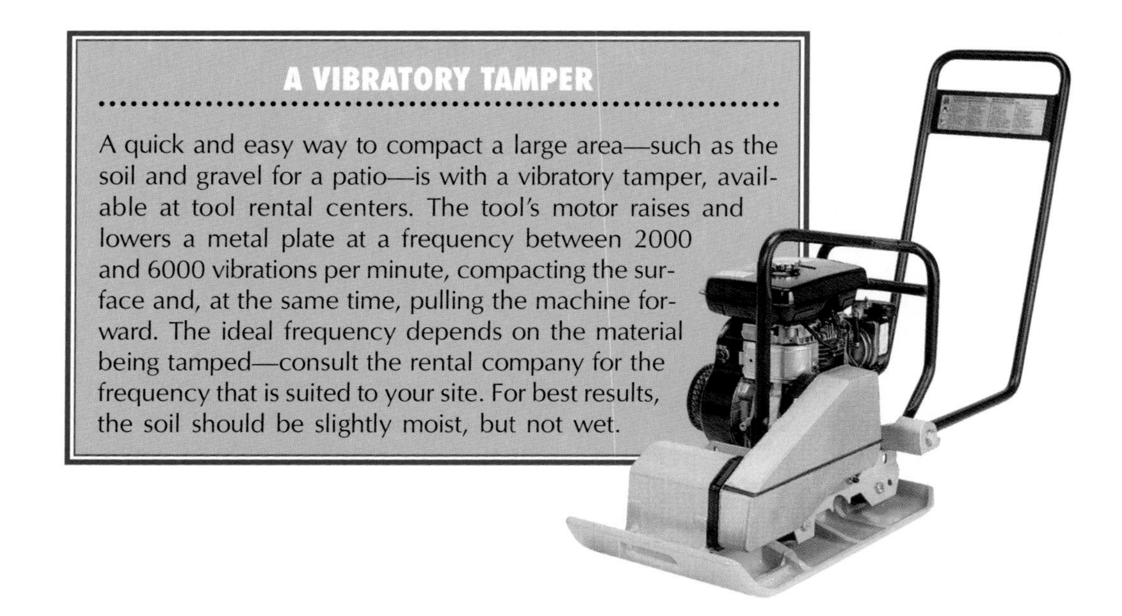

A VIBRATORY TAMPER

A quick and easy way to compact a large area—such as the soil and gravel for a patio—is with a vibratory tamper, available at tool rental centers. The tool's motor raises and lowers a metal plate at a frequency between 2000 and 6000 vibrations per minute, compacting the surface and, at the same time, pulling the machine forward. The ideal frequency depends on the material being tamped—consult the rental company for the frequency that is suited to your site. For best results, the soil should be slightly moist, but not wet.

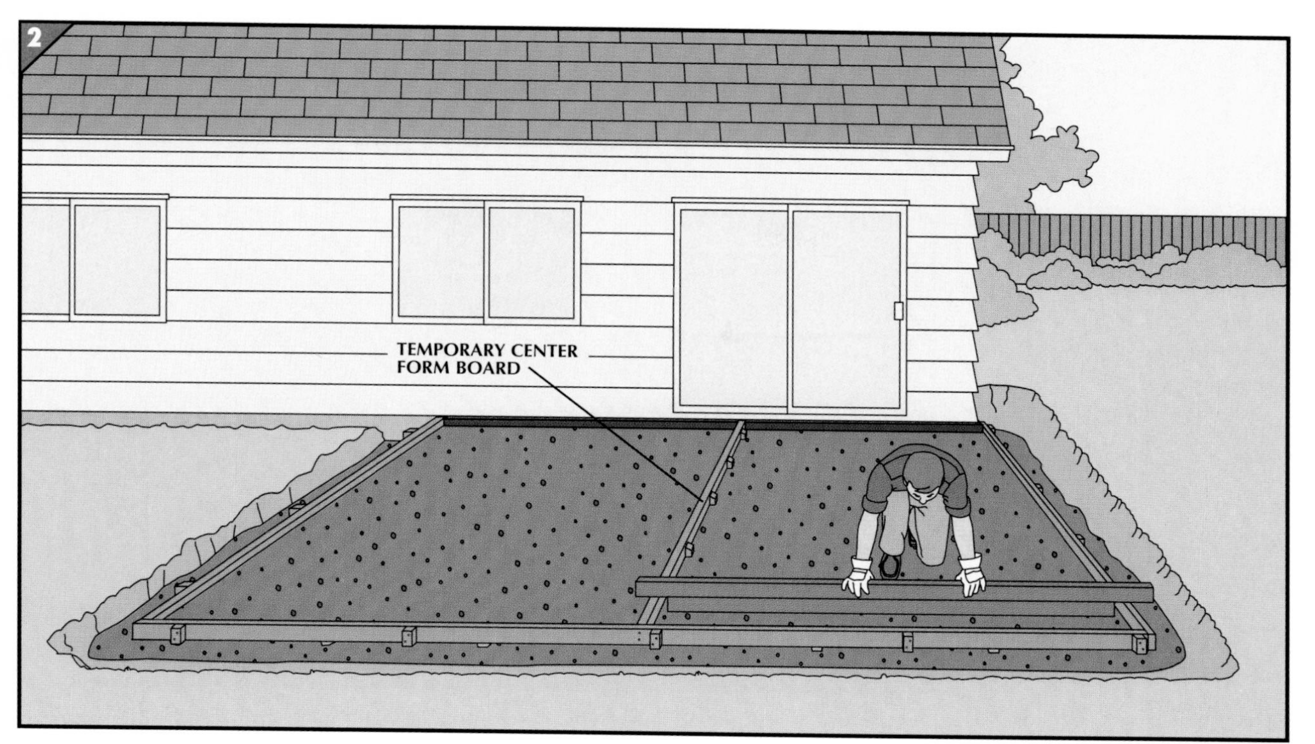

2. Preparing the bed.

◆ Place expansion-joint material against the house.

◆ Divide a large patio into sections no more than 10 feet wide by setting up temporary stakes and a center form board at a right angle to the house.

◆ Lay a gravel bed 4 inches deep and screed one section at a time *(page 46)*, resting the screed on the form boards *(above)*.

◆ Tamp the gravel.

3. Reinforcing with wire mesh.

◆ Start laying wire mesh 2 inches from one corner of the slab section, weighing down the end with concrete blocks.

◆ Walk backward, unrolling the mesh, and cut it off 2 inches from the form board at the opposite end of the section. Because the wire will curl, work with a helper to turn the mesh over and flatten it by walking on it.

◆ Lay subsequent strips the same way *(above)*, overlapping the mesh by one square.

◆ Tie the ends of overlapping strips together with wire.

◆ Before pouring the concrete, lift the wire mesh and support it 2 inches above the gravel bed with bricks or stones spaced 2 to 3 feet apart.

CAUTION *Be extremely careful when working with wire mesh—it can spring free and cause injury as it is being unrolled.*

4. Placing the concrete.
◆ Pour the concrete, level, and compact it one section at a time *(pages 47-48)*, leaving the center form board in place.
◆ Using planks to walk across the slab, remove the temporary center board and stakes, and shovel concrete into the divide *(right)*.
◆ Level and compact the concrete along the center with the back of the shovel.

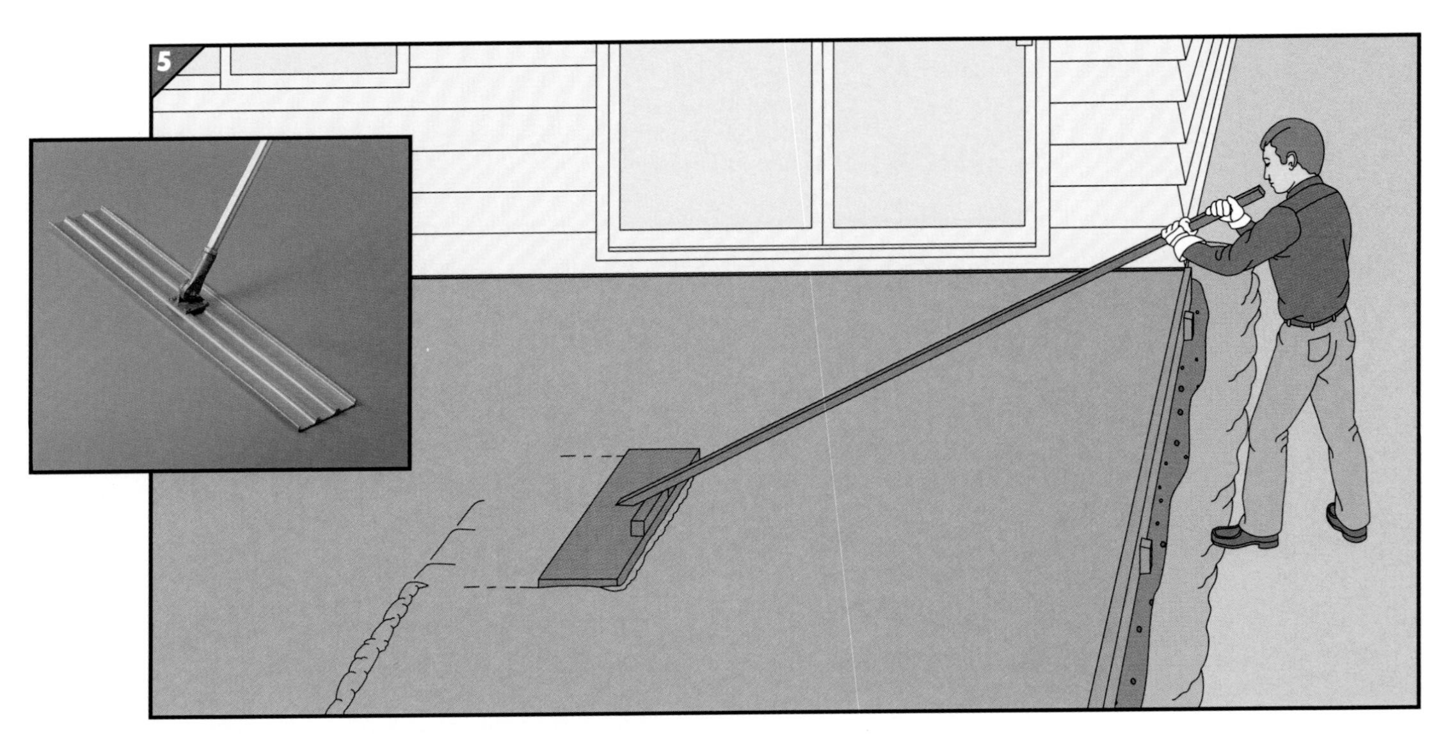

5. Finishing a large slab with a bull float.
◆ Make a bull float, starting with a 12-foot 2-by-2 as a handle; cut one end at an angle of about 10 degrees so the handle will rest flat on the float and the gripping end will be at eye level when the float is used. For the float end, use No. 8 wood screws to fasten a wood block to a 4-foot 1-by-12, then fasten the handle to both pieces, making sure the screws don't protrude through the bottom of the float. Alternatively, rent a commercial bull float *(photograph)*.
◆ Starting at one corner of the slab, push the float away from you across the surface; push the handle down to raise the leading edge of the float and prevent it from digging into the slab. Then pull the float back toward you with the float flat on the surface *(above)*.
◆ Float the slab strip by strip until you reach the opposite end, then repeat from the other side of the slab.
◆ Complete the finishing and curing steps described on pages 49 to 53.

TRICKS OF THE TRADE

Kneeboards to Extend Your Reach

An alternative to building or renting a bull float is to use kneeboards made from 1-by-2-foot pieces of $\frac{1}{2}$-inch plywood. Nail a 2-by-2 to each end of the plywood as handles. Work with two kneeboards, reaching as much of the surface as possible from one board and then moving the second one ahead of you. To finish the slab efficiently, alternate between the float and the steel trowel, finishing a section with both tools before moving on. You can use the second tool to support yourself as you lean out.

A PATIO WITH PERMANENT FORMS

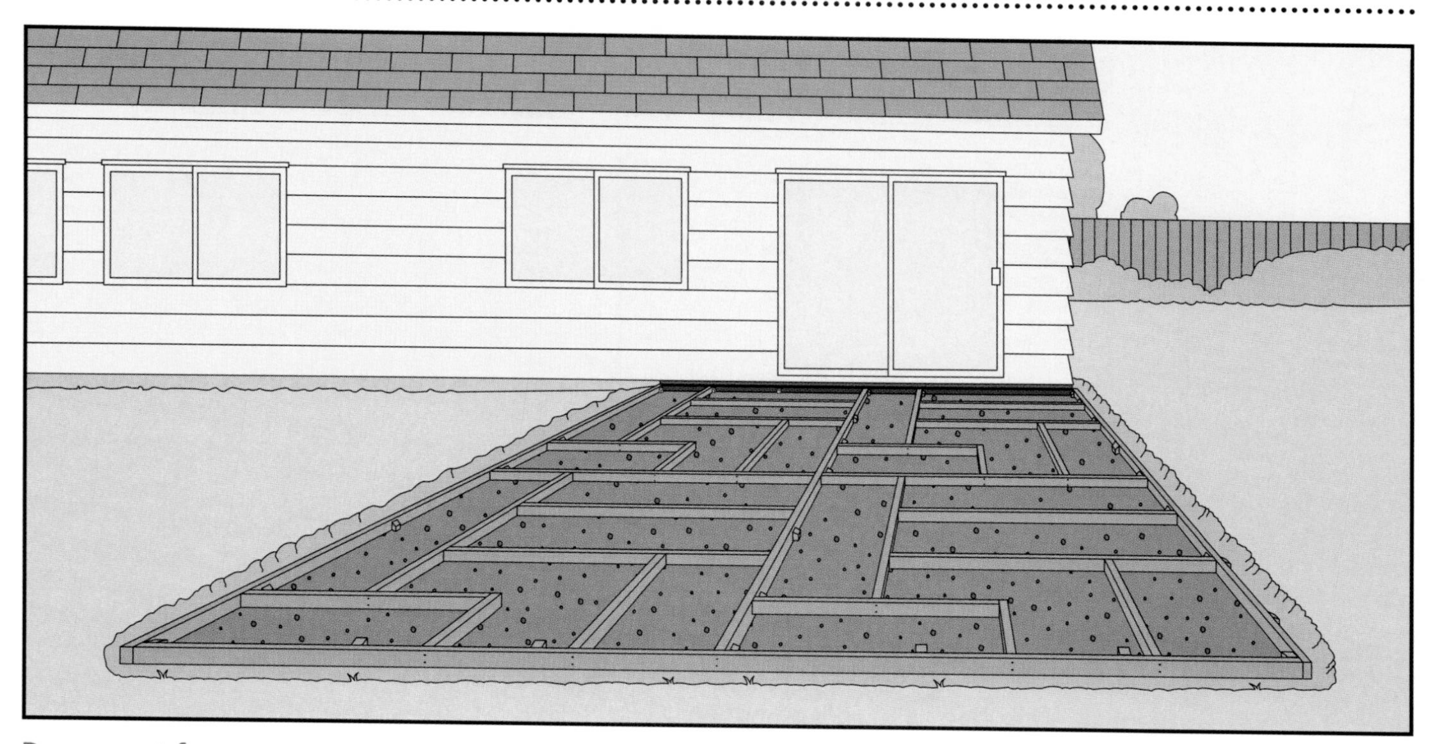

Permanent forms.
Forms that can be left in place permit you to divide a large patio into a grid of small sections that can be poured a few at a time —ideal if you wish to mix the concrete yourself and build the patio at your own pace. For the form boards and stakes, use pressure-treated lumber, or redwood or cedar treated with a clear wood sealer. Form boards can be laid out in any rectangular pattern, but sections less than 25 square feet are easiest to manage. For sections longer than 8 feet, cut control joints *(page 50, Step 2)* to divide the sections into areas no more than twice as long as they are wide.

1. Laying out the grid.

◆ Place the stakes for the perimeter forms inside the form boards, driving them 2 inches below the tops of the boards. Attach the perimeter forms to the stakes.

◆ Lay out the grid for the remaining forms with strings running from the bottom of the perimeter form boards. At each point where strings cross, drive an 8-inch-long 2-by-4 stake down to the level of the string *(right).*

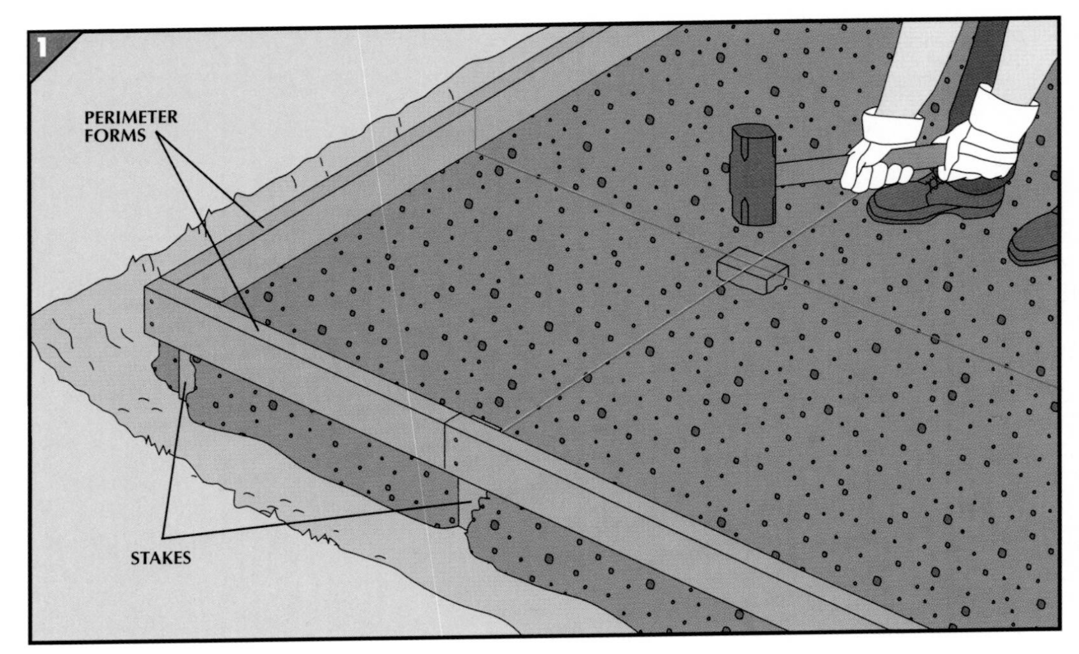

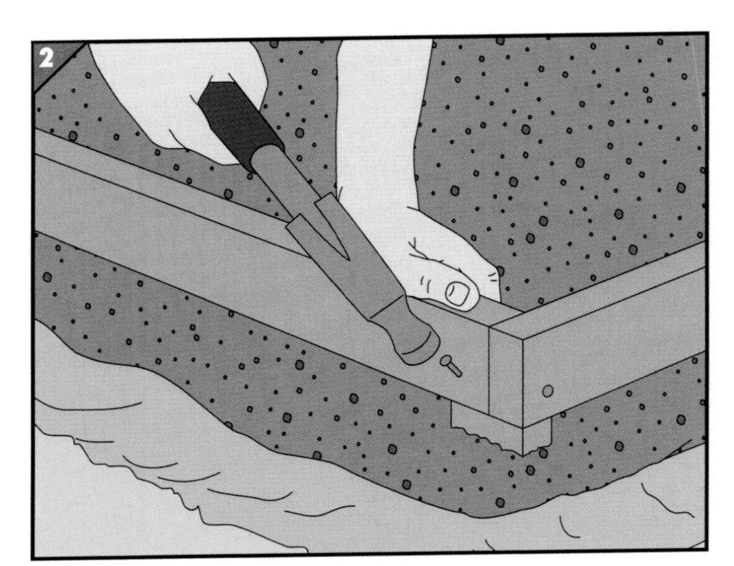

2. Securing forms to the stakes.

◆ Rest one form board on a stake and fasten it in place with a 3-inch galvanized common nail driven at an angle.

◆ Butt another board against the first and secure it with an angled nail *(left),* then nail through the first board into the end of the second.

3. Adding anchor nails.

To bind the concrete to the form boards, drive 3-inch galvanized nails into the sides of the perimeter boards midway between the top and bottom at 16-inch intervals. Hold a maul on the outside of the boards to prevent them from moving, and leave about half the nail sticking out. Repeat on the interior form boards, alternating the nailheads between the back and front of the boards *(right).*

Protect the tops of the boards with heavy-duty masking tape while pouring and finishing the concrete.

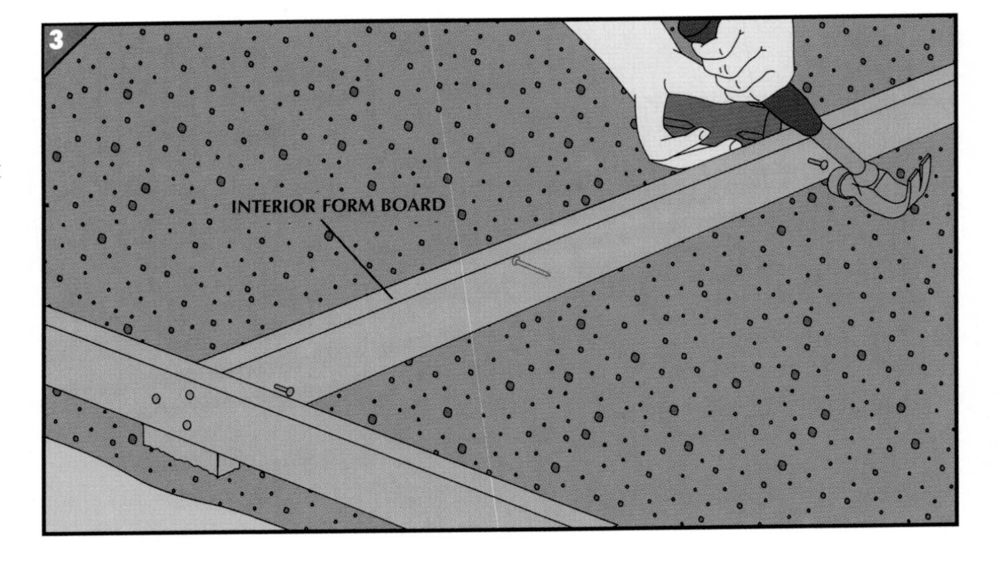

A Driveway that Lasts

A concrete driveway may cost more to build than an asphalt one, but the payoff is a longlasting, virtually maintenance-free surface. Although built with nearly the same tools and techniques as a walkway *(pages 36-53)*, it requires a curved access at the street; it may also be graded differently. If only cars will use the driveway, a 4-inch slab is sufficient. If heavy vehicles like oil-delivery trucks will drive over it, make the slab 6 inches thick.

Planning: For a single-car garage, make the driveway at least 10 feet wide; for a two-car garage, make it 21 feet wide at the garage (21 feet away from the garage, it can begin narrowing to 10 feet). On a scale drawing of the site, plot adjoining features such as the garage, plantings, and house; add the street access and a turning area if you need one. Check local codes for breaking through the sidewalk and curb.

Drainage: Check the codes for drainage requirements. If the garage is higher than the street, slope the driveway as you would a walkway *(pages 41-42)*. If the site is level and above the street, it's best to build a crowned driveway—raised 1 inch in the center—so water runs off both sides. But if you want water to drain to one side only, pitch the driveway in that direction as you would for a patio. Where the garage is below street level, make a concave driveway—one that slopes from the sides toward the middle, and install a drain where the driveway meets the garage entrance.

Along its length, slope the driveway no more than $1\frac{3}{4}$ inches per foot; otherwise, the bottom of the car may scrape against the surface. Make sure the driveway slopes from the sidewalk to the street.

TOOLS

Tape measure
Hammer
Maul
Shovel
Tin snips
2 x 4 screed
Darby or magnesium float
Jointer
Steel float

MATERIALS

Stakes
1 x 4 or 2 x 4 form boards
Gravel
Common nails ($\frac{1}{2}$", 3")
Expansion-joint material
Flashing metal

SAFETY TIPS

Wear a dust mask and goggles when mixing concrete. Protect your eyes with goggles when placing or screeding concrete. Also put on work gloves when working with concrete. When handling concrete directly, switch to rubber gloves.

CASTING A CONCRETE DRIVE

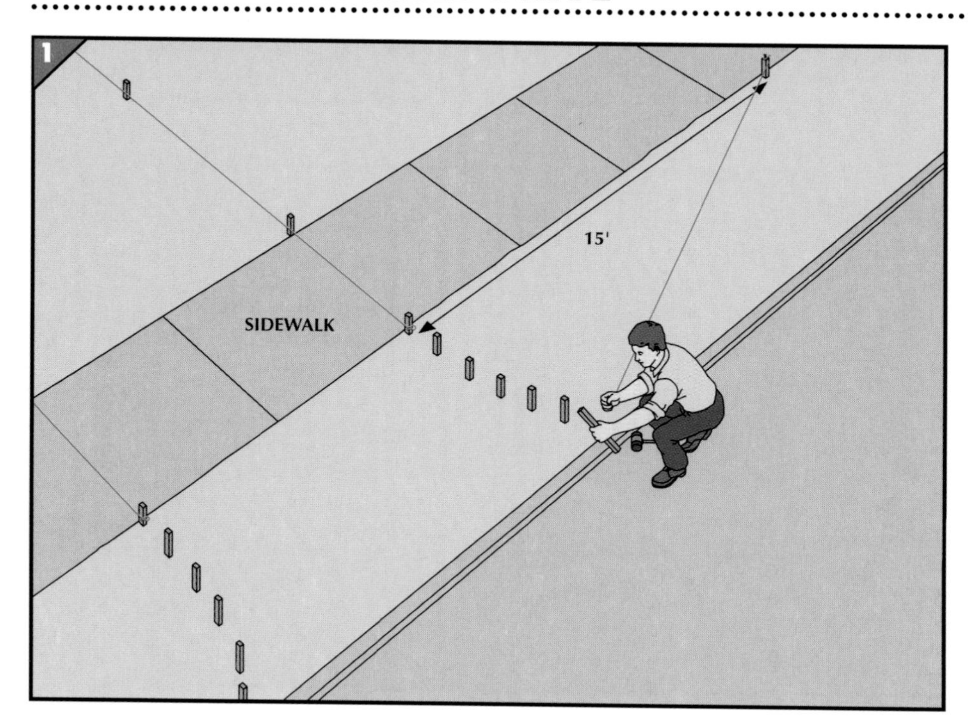

1. Staking the street access.

◆ Lay out and stake the straight part of the driveway as you would a walkway *(page 39, Step 1)*.

◆ Begin the curve of the driveway by first driving a stake at the outside edge of the sidewalk 15 feet from each side of the layout.

◆ Drive a nail into the stake's top end, and tie a 15-foot-long string to the nail.

◆ Use the end of the string as a guide to mark the curve from driveway to street. Drive a stake every foot from the end of the straight section to the edge of the street.

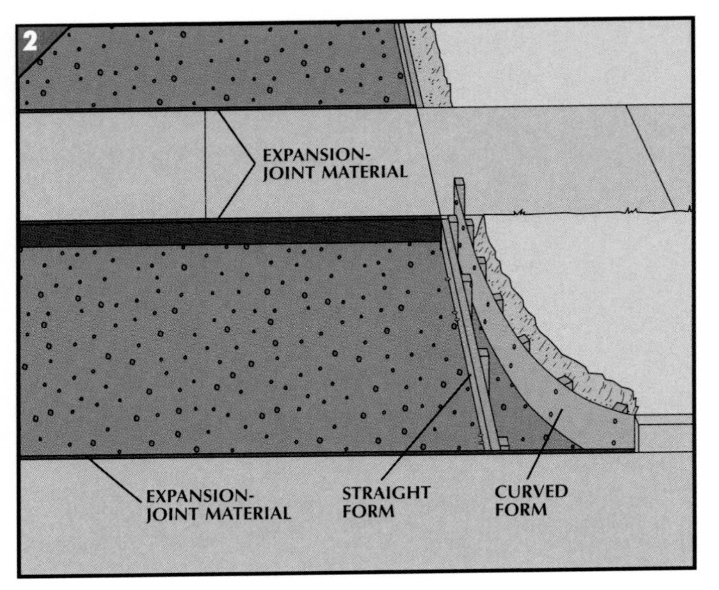

2. Forms for the street access.

◆ For a 4-inch-thick driveway, excavate the site to 8 inches below street level where the driveway will meet the street and 8 inches below the level of the sidewalk. Dig 2 inches deeper for a 6-inch-thick driveway. If you're grading the driveway to drain to one side, use the method on page 40 for sloping a walkway across its width.

◆ Set up form boards *(page 43)* out to the street with their top ends level with the sidewalk and their bottom ends level with the street *(left).*

◆ Form the curve plotted in Step 1 with flashing metal cut to the height of the curb plus the thickness of the concrete. Starting flush with the curb, attach the metal to the stakes with $\frac{1}{2}$-inch nails; continue to the sidewalk as shown.

◆ Lay a gravel drainage bed *(page 46)* and set expansion-joint material *(page 46)* against both edges of the sidewalk, and against the edge of the street.

3. Finishing the street access.

◆ Place the concrete *(pages 47-48, Steps 1 and 2).*

◆ If you've graded the driveway to drain to one side only, screed the concrete with a straight 2-by-4 *(page 48, Step 3)*; otherwise, use a curved screed *(below).*

◆ At the street, screed the concrete between the straight boards, then pull out the straight boards.

◆ Fill the space at the curved forms with concrete.

◆ Shape the sides of the slab at the street with your hands, tapering it from the top of the curb into the driveway's curve and sloping it down to the flat surface of the driveway *(right).* Smooth this area with a darby or float.

◆ Finish the rest of the slab *(pages 49-53)*; you'll need to cut control joints across the width of the driveway and, for driveways over 10 feet wide, up the center. Use a small float and trowel for the edges at the street.

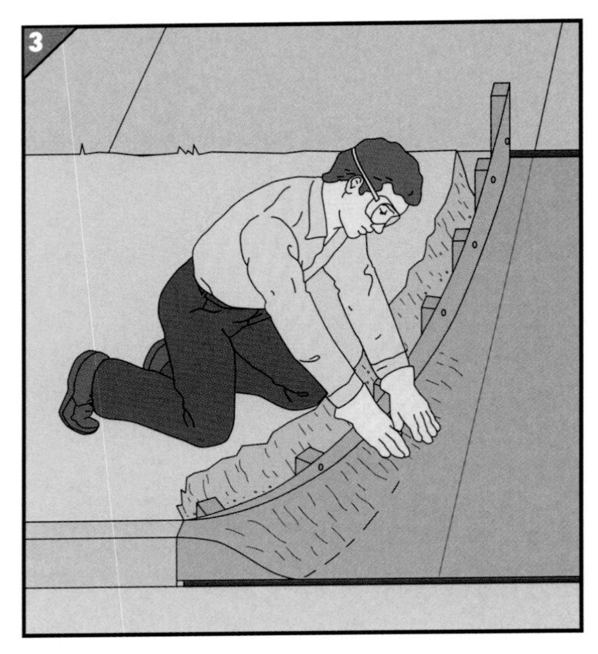

TRICKS OF THE TRADE

Curved Screeds

To smooth a crowned or concave driveway, make a curved screed: Cut a 2-by-6 and a 2-by-4 to the width of the driveway and place them flat on the ground alongside each other. For a crowned driveway, nail a piece of scrap wood across the center of the boards on both sides. At each end, insert $\frac{3}{4}$-by-$1\frac{1}{2}$-inch shims between the boards so the ends bow out, then nail scrap wood at each end to hold the bow. For a concave driveway, fasten the ends of the boards together first, place a shim at the center, then secure the center.

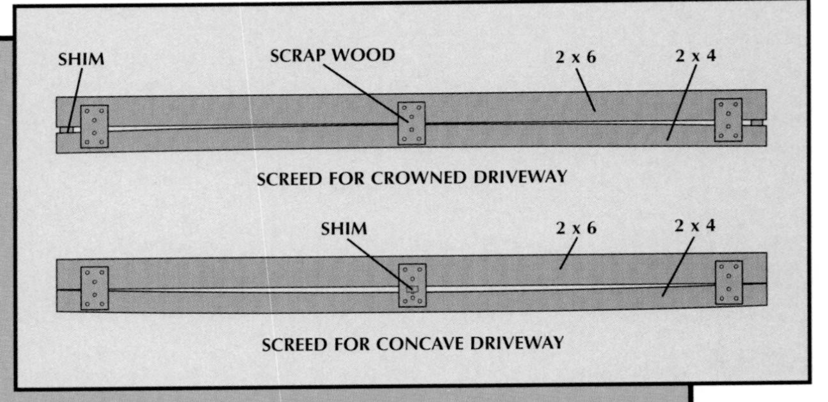

Sturdy Footings for Strong Walls

Below the base of a garden wall, usually concealed by earth or sod, lies a vital structural element—the footing. This takes the form of a long, narrow, flat-topped concrete slab that supports all the weight of the structure above.

Footing Design: Footings for the low freestanding brick, block, and stone walls described in this book do not differ much from an ordinary concrete slab except that they must be built below the frost line on solid, undisturbed soil with no drainage bed. Footings are twice the width of the wall and about 8 inches thick; using 2-by-8 form boards will result in a $7\frac{1}{2}$-inch-thick slab, which is acceptable. Locate the base of the footing at least $\frac{3}{4}$ inch below the surface; and in freezing climates, 6 inches below the frost line—this is required by code in some areas, and protects the wall from cracking as the ground freezes and thaws. The wall itself will then start below ground. In very cold climates, consider renting a power trencher, as you may have to excavate as deep as 6 feet. (The illustrations on the following pages represent a footing located just below the surface, in a nonfreezing climate.)

Some local codes, particularly in earthquake areas, require that footings be strengthened with steel reinforcing bars (rebars) to control cracking *(page 64)*. And if the wall itself needs to be reinforced *(pages 82 and 117)*, vertical rebars must be embedded in the freshly cast footing. To lay out footings for a wall with corners, use the method on page 88.

Building Forms: On level ground with firm soil, wooden forms may not be needed: a straight-sided trench will serve. Otherwise, build a simple form as shown below.

TOOLS

Shovel	2 x 4 screed
Hammer	Darby or float
Handsaw	

MATERIALS

Stakes	1 x 2s
2 x 8 form boards	Common nails ($2\frac{1}{2}$", 3")

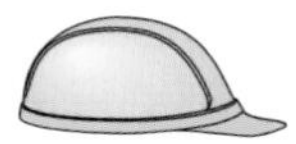

SAFETY TIPS

Concrete is caustic—wear gloves when working with it. If you are mixing concrete, put on a dust mask and goggles. Protect your eyes with goggles when pouring and screeding concrete, or when nailing or driving in stakes.

1. Building the form.

◆ Dig a trench about 2 feet wider than the footing, with the bottom of the trench 6 inches below the frost line.

◆ Build forms with stakes and 2-by-8 form boards as described for a concrete walkway *(pages 43-44)*, but make the trench and forms absolutely level. Do not follow a sloping grade or introduce a slope for drainage.

◆ Fasten 1-by-2 cleats called spreaders across the tops of the form boards every 3 to 4 feet with $2\frac{1}{2}$-inch common nails.

2. Pouring the concrete.

◆ Mix and pour concrete for the footing as you would for an ordinary slab *(pages 47-48, Steps 1 and 2)*, making sure that the spaces immediately below the spreaders are completely filled.
◆ Remove the spreaders carefully without disturbing the semi-fluid concrete below them.
◆ With a straight length of 2-by-4 to serve as a screed, level the concrete flush with the top of the forms *(left)*.
◆ Smooth the concrete with a darby or float and cure the concrete *(page 53)*.

A REINFORCED FOOTING

In some areas, codes require footings to be strengthened by rebars; the exact size and number of rebars are specified in the code. Build forms for the footing and, before fastening spreaders across the tops of the forms, run rebars from one end of the forms to the other. These steel rods come in 20-foot lengths, typically $\frac{3}{8}$ to $\frac{1}{2}$ inch thick. Wire them to rocks or bricks so that they lie about half of the footing depth above the bottom of the trench. Overlap rebars about 18 inches and fasten them with wire.

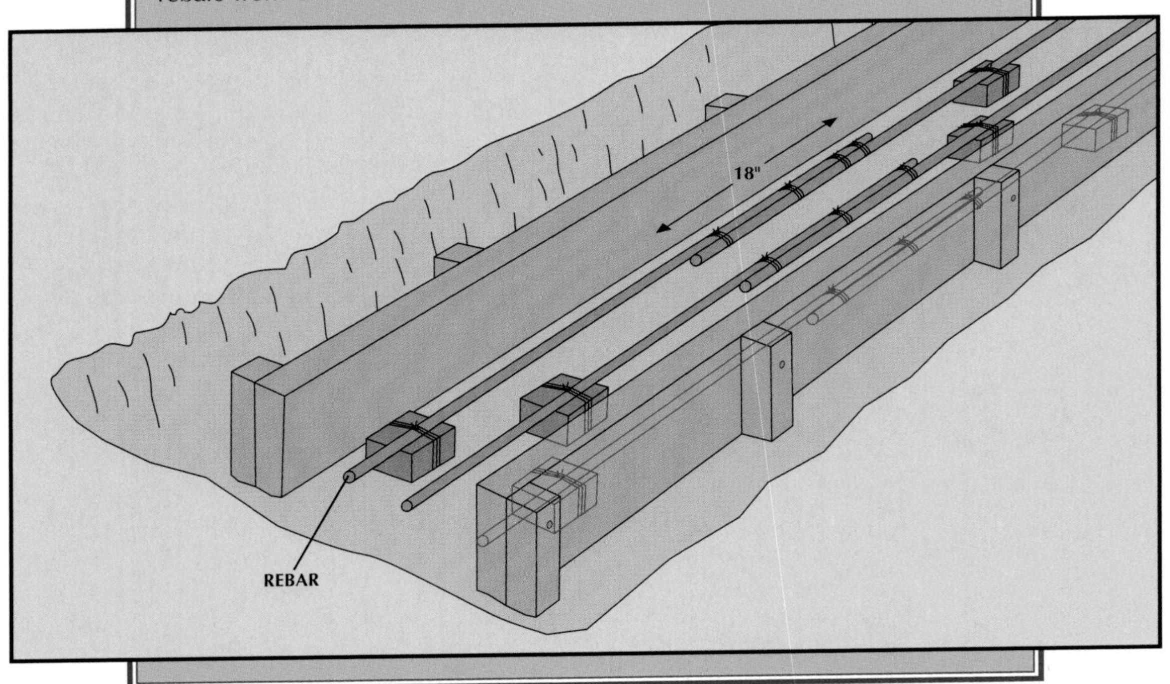

Rugged Steps of Cast Concrete

When you need to replace worn-out exterior wood stairs, or provide new ones for an added doorway, cast concrete is a good choice: Concrete steps are fairly simple to make and, once built, are nearly indestructible. The method shown here is suitable for a stairway a maximum of about $2\frac{1}{2}$ feet high and not attached to a building.

Step Proportions: The combined depth of the tread (or top of the step) and the height of the riser (the vertical section) is governed by code; they should typically total about 18 inches. As a practical matter, risers should be 6 to 8 inches high, and they should be combined with treads of 12 to 10 inches. Determine the dimensions for your project by measuring and calculating *(page 66, Step 2)*.

Laying Out Side Forms: Make the forms for the sides from plywood sheets. For comfort and weather resistance, gently angle the steps' profile, with the treads sloping down slightly and the risers leaning forward a bit *(page 66)*. Since the sides of the steps will be visible, brush form-release agent onto the inside surfaces of the form boards—it will enable the forms to come off cleanly.

TOOLS

- Tape measure
- Carpenter's square
- Protractor
- Level
- Shovel
- Saber saw
- Hammer
- Maul
- Tamper
- Paintbrush
- Square shovel
- Flat spade

MATERIALS

- 2 x 4s
- 2 x 8s or 2 x 10s
- $\frac{3}{4}$" plywood
- Gravel
- Form-release agent
- Common nails (2", $3\frac{1}{2}$")
- Expansion-joint material

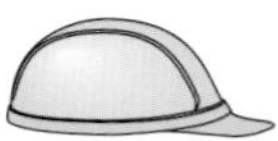

SAFETY TIPS

Wet concrete is caustic—wear gloves when working with it. If you are mixing concrete, wear a dust mask and goggles. Protect your eyes with goggles when pouring or screeding concrete and when nailing or driving in stakes.

1. Finding the height.

◆ Dig down as necessary to level the ground in the area approximately 6 feet in front of the doorframe.

◆ Measure from the ground to the underside of the sill *(right)*: This will be the total rise—the vertical distance that the stairs will fill when completed.

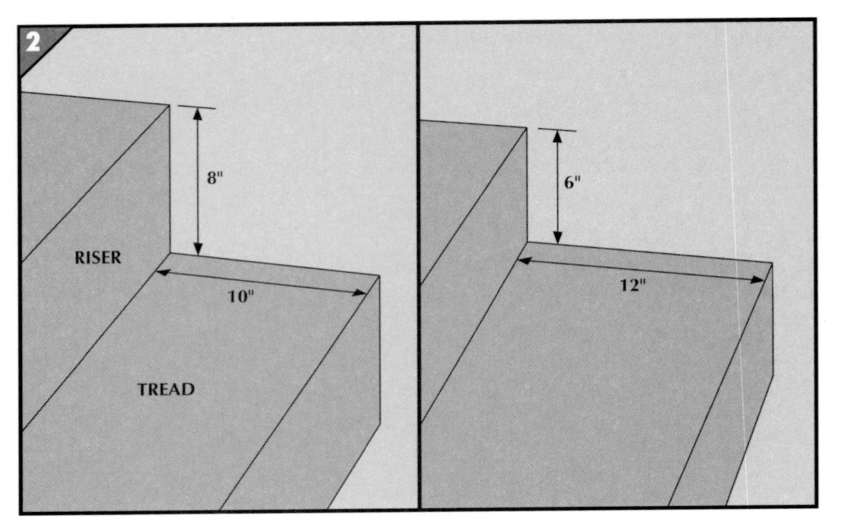

2. Figuring dimensions.

Make the staircase at least 6 inches wider than the door opening. The uppermost step, or landing, should be at least 3 feet deep to provide an area for entering and leaving safely.

◆ To find the height of each riser, divide the rise *(page 65, Step 1)* by the number of steps you wish to build. Subtract the riser height from 18 inches to obtain the tread width. Thus, if the rise is 24 inches, you could make three steps, each with an 8-inch riser and 10-inch tread; or, four gentle steps, each with a 6-inch riser and 12-inch tread.

◆ From the dimensions you obtain, draw a plan and estimate materials *(page 36)*.

3. Cutting the form sides.

◆ Lay out the sides of the forms on a sheet of $\frac{3}{4}$-inch plywood, marking the dimensions from Step 2. Make the height of the forms 12 inches more than the rise, to provide room for 6 inches of concrete and 6 inches of gravel below ground level.

◆ Draw risers and treads at right angles with a carpenter's square.

◆ Slope the landing and treads downward to provide drainage, pitching them $\frac{1}{4}$ inch for each horizontal foot; then slope the risers 15 degrees inward *(dashed lines)*.

◆ Cut the form with a saber saw, then use it as a template to cut the second one.

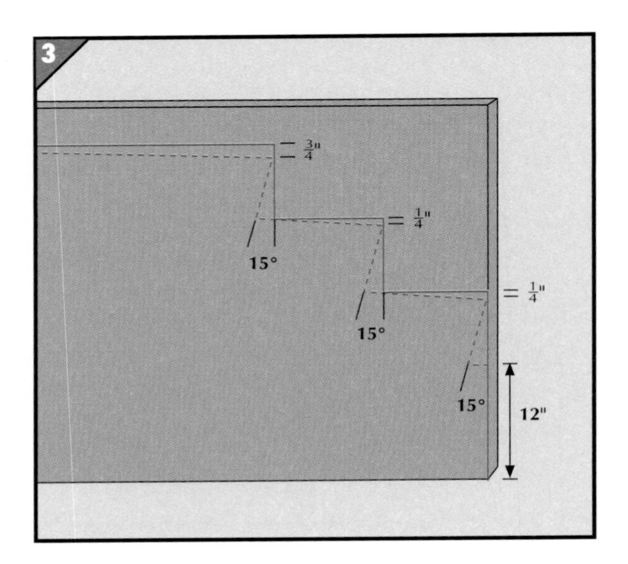

4. Setting up the form sides.

◆ Dig a hole 12 inches deep and about 1 inch larger all around than the steps. Compact the ground with a cast-iron tamper.

◆ Nail 2-by-4 supports to the outsides of the form sides with 2-inch common nails.

◆ Place the form sides against the sides of the hole, $\frac{1}{2}$ inch away from the house wall, checking with a carpenter's square to make sure they are at right angles to the house, and with a level for plumb.

◆ Drive 2-by-4 stakes at an angle at least 8 inches into the ground about 18 inches away from the hole.

◆ Nail 2-by-4 braces between the form sides and the stakes with $3\frac{1}{2}$-inch nails.

◆ Pour in 6 inches of gravel and tamp it down.

5. Completing the form.

◆ Cut lengths of 2-by-8 or 2-by-10 to the dimensions of the risers. To enable you to reach and smooth the concrete over the entire surface of the tread with a trowel, bevel the bottom edges *(inset).*
◆ Nail the riser boards to the form sides with 2-inch common nails.
◆ Cut a piece of asphalt-impregnated expansion-joint material to fit against the house *(right).*
◆ Brush form-release agent onto the inside surfaces of the forms to prevent concrete from sticking to them.

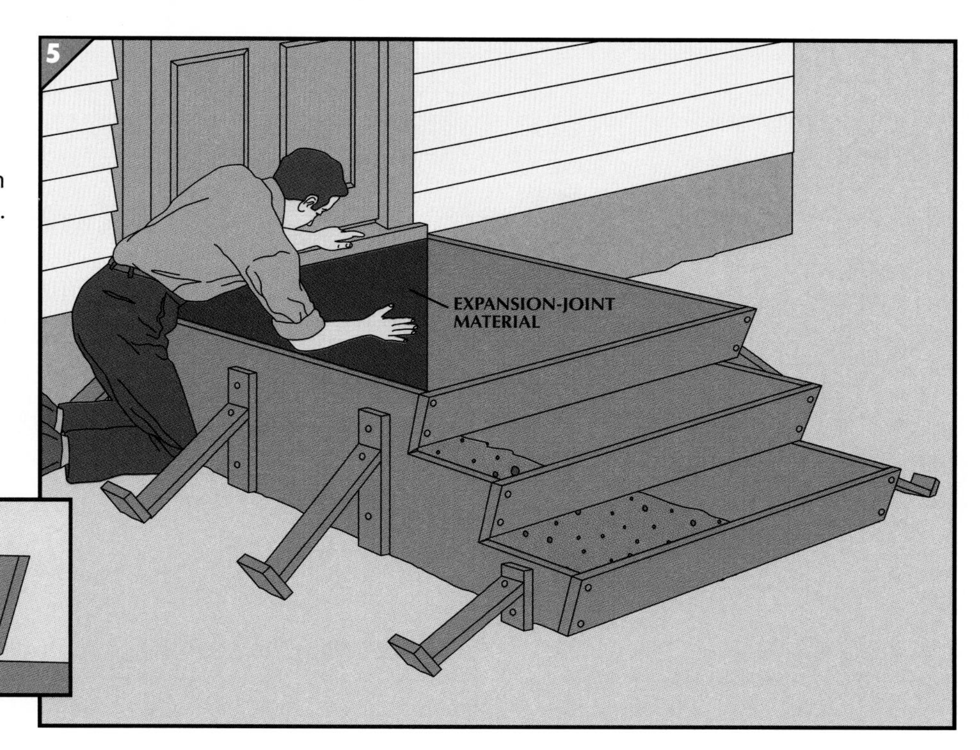

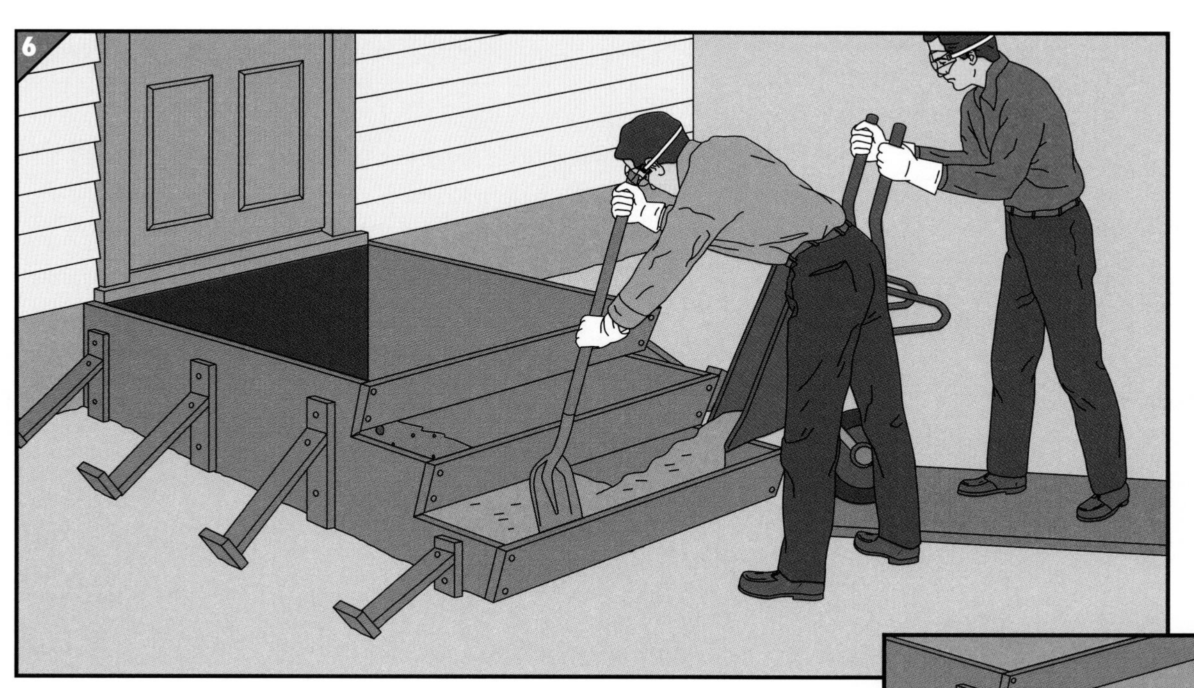

6. Placing the concrete.

◆ Pour concrete directly into the form, starting at the lowest step *(above)* and overfilling the form slightly; then work your way upward to fill the other steps. If the riser boards bulge out, brace them with 2-by-4s *(inset).*
◆ Drive square shovels into the corners to eliminate air pockets.
◆ When the pouring is completed, clear the edges of the forms with a flat spade *(page 48, Step 2).*
◆ Screed and smooth the steps *(pages 48-49, Steps 3 and 4).*
◆ Once the bleed water has evaporated, finish and cure the concrete *(pages 49-53)*, omitting the control joints.
◆ After the concrete has cured, carefully remove the forms. If necessary, patch any chips or holes in the concrete *(pages 19-24).*

3 Brick, Block, Tile, and Stone

Few building materials can match brick, concrete block, tile, and stone for strength and durability. They are versatile as well because of their many shapes, sizes, colors, and textures. Easier to work with than poured concrete, they yield attractive and longlasting structures such as walls, patios, steps, and backyard barbecues.

A mason's line for a straight row of bricks →

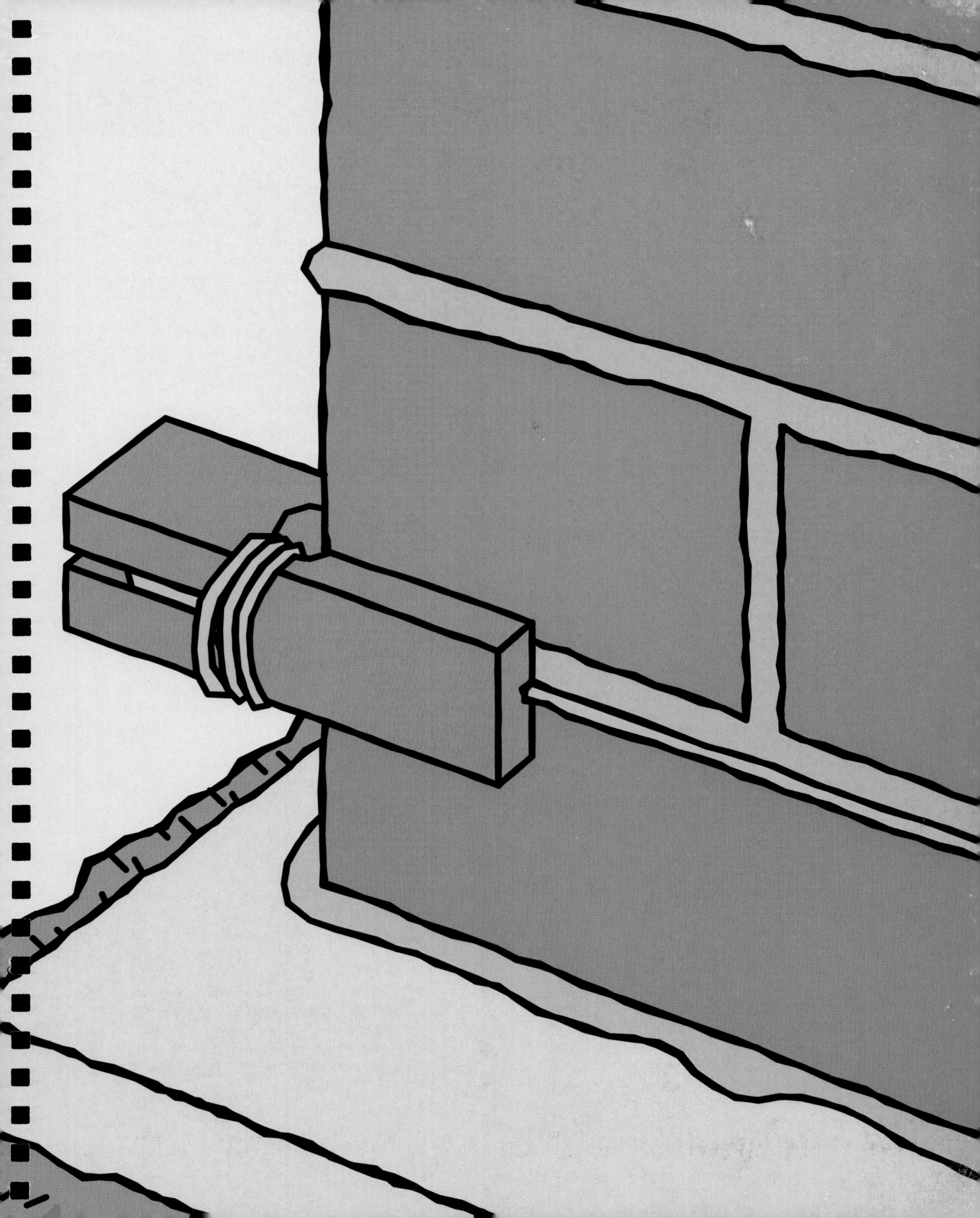

Choosing Bricks

Bricks are ideal building materials—compact and easy to handle yet strong and durable. Available in countless shapes, sizes, colors, and textures, most fit into three categories: building, face, and paving bricks.

Varieties of Brick: Generally red in color, building bricks are economically priced and suitable for outdoor use. However, they have a less finished appearance than face bricks, which range from white to purplish black, and from rough-textured to glassy-smooth. Face bricks make distinctive walls, steps, and barbecues, but most are not suitable for walkways or patios. There, you will want to use paving bricks, which have one smooth, less absorbent surface, and are shallower (sometimes only $\frac{1}{2}$ inch thick) than other types. Paving bricks come in assorted colors—mostly shades of red—and in a variety of shapes. Precast concrete pavers come in a variety of shapes and can be laid in the same way as paving brick.

Some bricks are solid, and are either flat on all surfaces or have a depression and the manufacturer's imprint on one side. Other bricks have rectangular or round holes, called cores, running through them. Cored bricks yield stronger walls because some of the mortar runs down into the holes; the tops of walls built with cored bricks must be covered with solid bricks. Paving bricks do not have holes, but you can use cored building bricks for paving where your pattern calls for bricks set on their sides or ends.

Weather Resistance: All three types of bricks are rated according to their resistance to frost. Bricks rated SW or SX can withstand severe weather; MW or MX bricks should be used only in climates not subject to freezing temperatures; NW or NX bricks are for indoor use only.

Estimating and Ordering: Bricks are usually least costly if you buy them in prepackaged cubes of 500 to 1,000 units. Buying by the cube ensures that all the bricks are about the same color and size.

To estimate how many bricks you will need for a project, first calculate the total area to be covered in square feet. Divide irregular areas into squares or rectangles and add up the areas of the segments. For a double-layer wall, be sure to allow for both layers. Calculate the number of bricks needed per square foot as follows: For unmortared paving, multiply the length by width (in inches) of the brick surface that will be visible; divide the result into 144 for the number of bricks per square foot. For mortared applications, add $\frac{1}{2}$ inch to each dimension of the brick before multiplying, then divide the result into 144. Next multiply the number of square feet by the number of bricks per square foot. Finally, add at least 5 percent for cutting and breakage.

To estimate mortar requirements for paving, figure 8 cubic feet for 100 square feet of bricks. For a double-layer wall, figure about 20 cubic feet of mortar for each 100 square feet of bricks in both layers; for brick-veneered steps, about 12 cubic feet per 100 square feet. A $\frac{1}{2}$-inch-thick mortar bed for paving, walls, or steps requires about 5 cubic feet of mortar for each 100 square feet.

Before the bricks arrive, clear a delivery space close to the street or driveway and as near as possible to the work site. Build a wood platform of boards laid across 2-by-4s for the bricks. Cover the bricks with plastic sheeting.

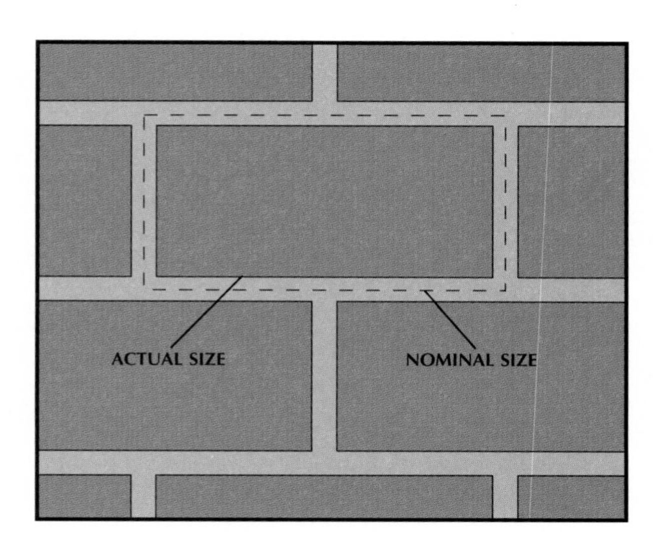

Brick sizes.
Brick dealers and masons sometimes refer to a brick by its nominal rather than its actual size. A brick's nominal size is its dimensions as measured after it has been mortared into a wall, with a $\frac{3}{8}$ or $\frac{1}{2}$ inch mortar joint included in the measurement. The standard nominal size of a brick is 8 by 4 by $2\frac{2}{3}$ inches. The sizes represent only an average, because bricks from the same lot may vary in size as much as 3 percent.

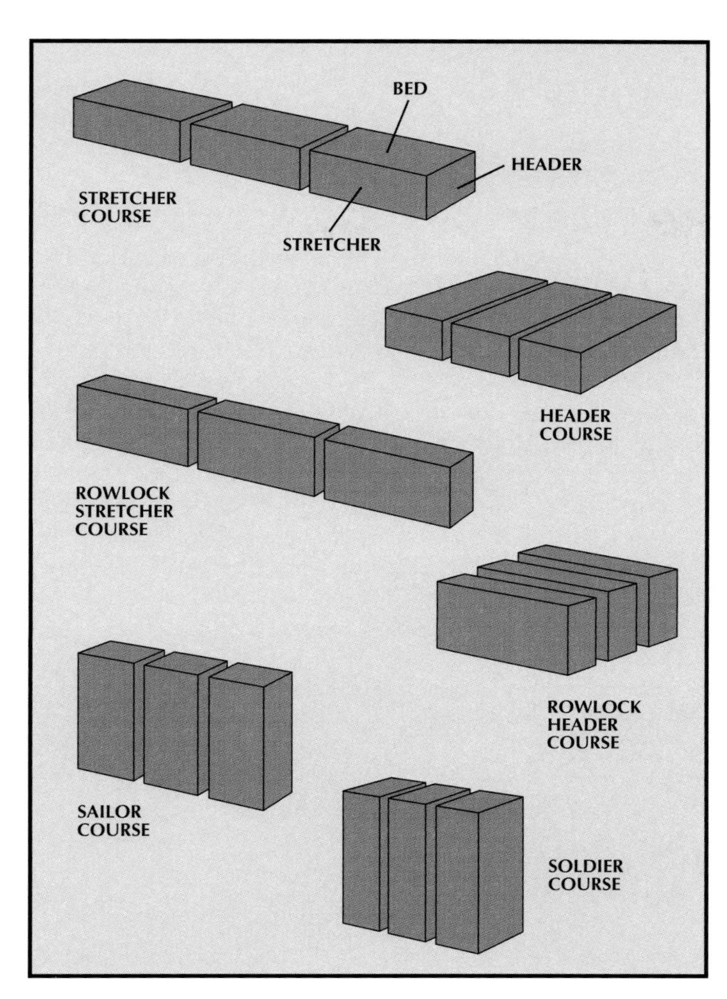

Bricklayers' jargon. The colorful names that masons apply to positions of bricks and arrangements of rows come mainly from terms for brick surfaces. The long sides are called stretchers, the ends headers, and the tops and bottoms beds. When bricks are laid on beds with headers abutting, the stretchers are exposed and the row is a stretcher course. When they are laid on beds with stretchers abutting, the headers are exposed and the row is a header course. Bricks laid on stretchers form a rowlock stretcher course when headers abut, a rowlock header course when beds abut. Bricks laid on headers form a sailor course when stretchers abut, a soldier course when beds abut.

SEVEN STYLES OF FACE BRICK

Bricks are available in a variety of surface colors and textures. Shown here is a sample of common types of face brick. Smooth, wire-cut, rug, and sand-struck textures are created as the brick is formed. Flashing is a method used in the firing of the bricks to add streaks of color. Bricks can be made from a variety of colors of clay; colored coatings can also be added. Tumbled bricks are new bricks given a used appearance.

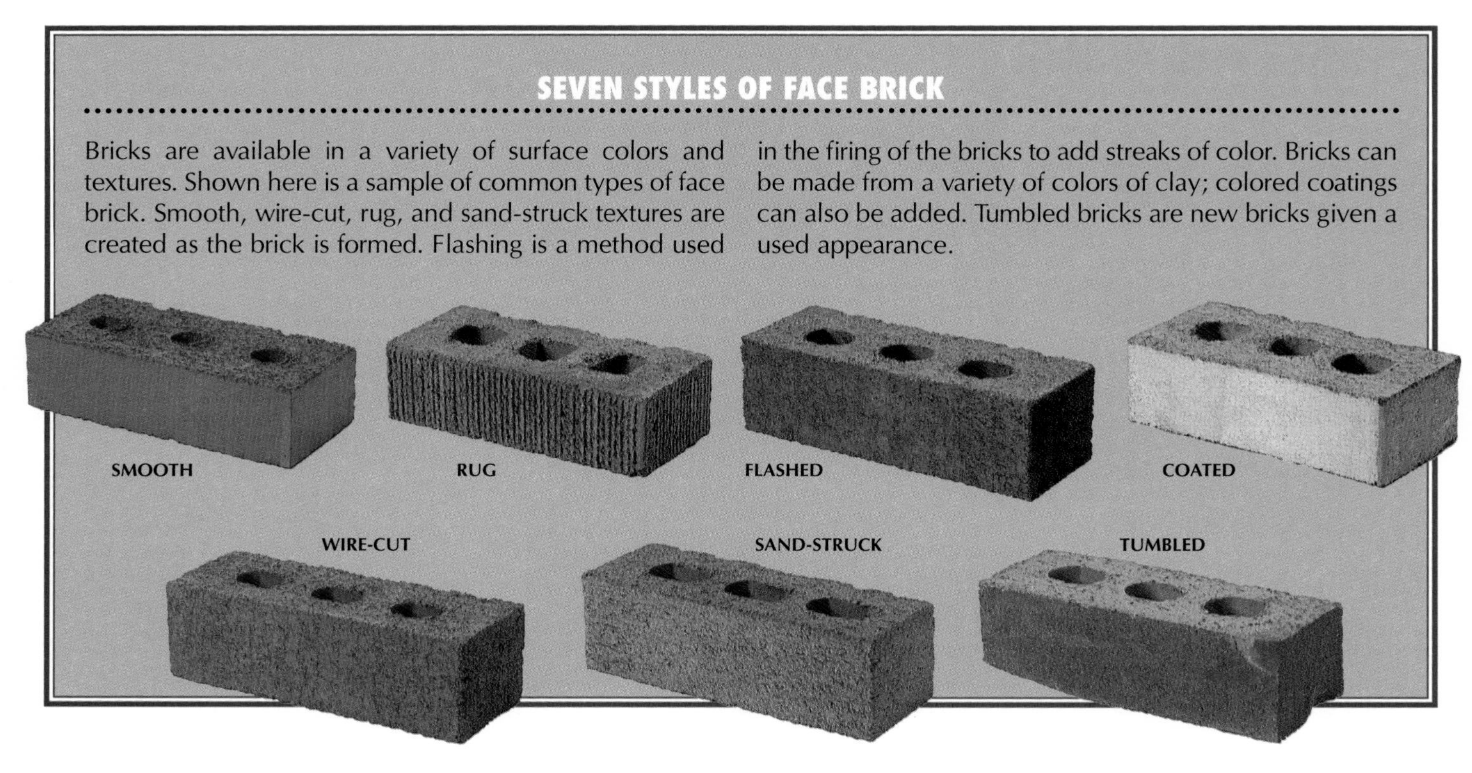

Bricks For Patios and Walkways

The chief advantages of brick paving are its handsome appearance and ease of construction. Set into a bed of sand or laid in mortar over a concrete foundation, bricks form an attractive and rugged walkway or patio *(pages 74-79)*.

Popular Patterns: Six classic patterns for laying bricks are shown below. Slight adjustments will yield the variations opposite. Each pattern requires the same number of bricks for a given area, but laying bricks on edge rather than flat requires 50 percent more. For some patterns, you'll need to cut bricks *(page 17)*.

A plain pattern such as jack-on-jack serves well for small spaces. The more intricate herringbone or basket weave are better for larger areas since several pattern repeats are required before the design is revealed. Bricks for all but the jack-on-jack and running bond patterns must be twice as long as they are wide—nominally for mortar joints; actually if mortarless *(page 70)*.

To make sure your pattern choice will work out as you imagined, sketch your walkway or patio area to scale before ordering any bricks. Then cut out miniature paper or cardboard rectangles of the size of your bricks and experiment with different combinations on the drawing.

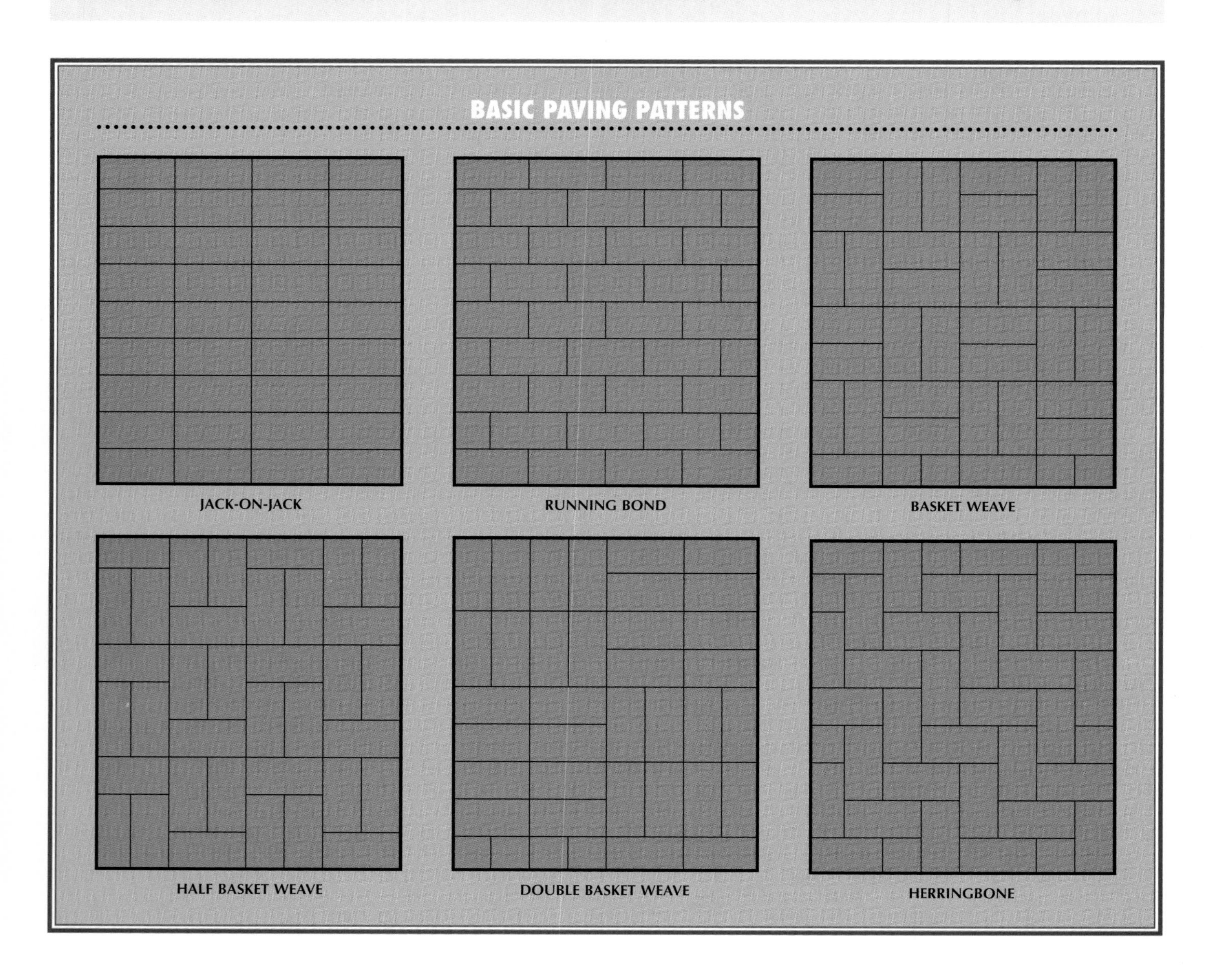

PATTERN VARIATIONS

Shifting a pattern diagonally.
Arrange your pattern at a diagonal to the edges of your path or patio rather than at right angles. Cut bricks at an angle to fill the ends of each course.

Standing bricks on edge.
By setting all the bricks in a pattern on their sides, you get a lighter look than when the bricks are laid flat. To use bricks edgewise in a basket-weave or herringbone pattern, select bricks whose thickness is equal to approximately one-third of their length—$2\frac{2}{3}$ by 8 inches, for example.

Shifting the joints.
A variation on running bond offsets the ends of bricks in successive rows by one-third or one-quarter the brick length, rather than one-half.

Changing directions.
When two jack-on-jack patterns are set at right angles to each other, a simple pattern becomes more visually compelling.

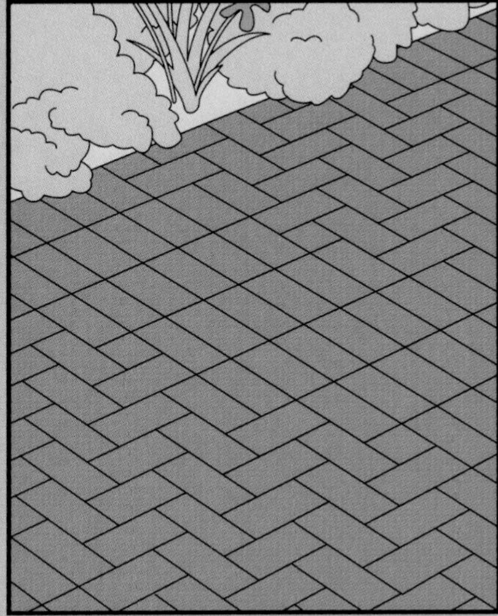

Combining patterns.
For large areas, you can combine two or three patterns. Here, a herringbone alternates with a plain jack-on-jack. Keep in mind that blending two fancy patterns may be distracting.

Gridding patterns.
For structural strength, build wood grids into the walkway or patio. The patterns within the grids may be identical *(above)* or contrasting.

Paving a Path or Patio

A patio or walkway can be built simply by laying paving bricks or concrete pavers unmortared in a bed of sand *(pages 76-77)*; but for a more permanent construction, your best bet is to set the units in mortar over a concrete slab *(pages 78-79)*.

Unmortared Paving: Easy to build and with a rustic charm, this style of paving allows you, with a minimum of effort, to salvage the brick later for reuse. In harsh climates, however, the surface may need to be leveled after a few years.

Mortared Paving: For this method, bricks rated SX *(page 70)* are laid over a concrete slab. The slab may be an existing sidewalk or patio, or you can pour one yourself using the techniques on pages 36 to 60, making sure to reinforce it with mesh *(page 57)*. As with tile *(page 91)*, line up mortar joints with control joints in the slab, cutting bricks as necessary.

Edgings: Bricks laid in sand should be enclosed in edging made of brick or of other materials such as wood or concrete *(opposite)*. For a decorative effect, combine wood edging with wood dividers set directly on the paving bed.

TOOLS

Tape measure
Maul
Shovel
Garden trowel
Tamper
2 x 4 screed
Mason's level
Stiff brush
Joint filler

MATERIALS

Stakes
String
Paving bricks
Wood strips (1" thick)
Concrete sand
Masonry sand
Edging boards
Common nails (3")
Mortar ingredients (Portland cement, masonry sand, hydrated lime)
1 x 8, 2 x 4 for screed
Muriatic acid

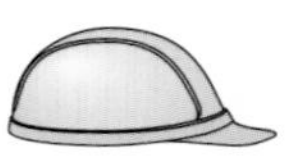

SAFETY TIPS

Mortar is caustic—wear gloves when working with it and goggles and a dust mask when mixing it. Gloves also protect your hands from the rough edges of bricks, and hard-toed shoes prevent injury from dropped or falling bricks. Put on goggles and rubber gloves to work with muriatic acid.

PLANNING THE LAYOUT

A dry run.

To establish the spacing between bricks and the dimensions of the paving bed and edging trench to be dug, you need to lay out a dry run of bricks.

◆ With stakes and strings, mark the area to be paved as you would for a concrete slab *(page 55)*.

◆ Enclose the area with edging bricks set on end up against the strings.

◆ Place paving bricks within the edging. For a simple pattern like a running bond, you can save time by laying out the sides and omitting the middle section.

If you are using concrete pavers *(photograph)*, place them as you would bricks, following the pattern dictated by their shape.

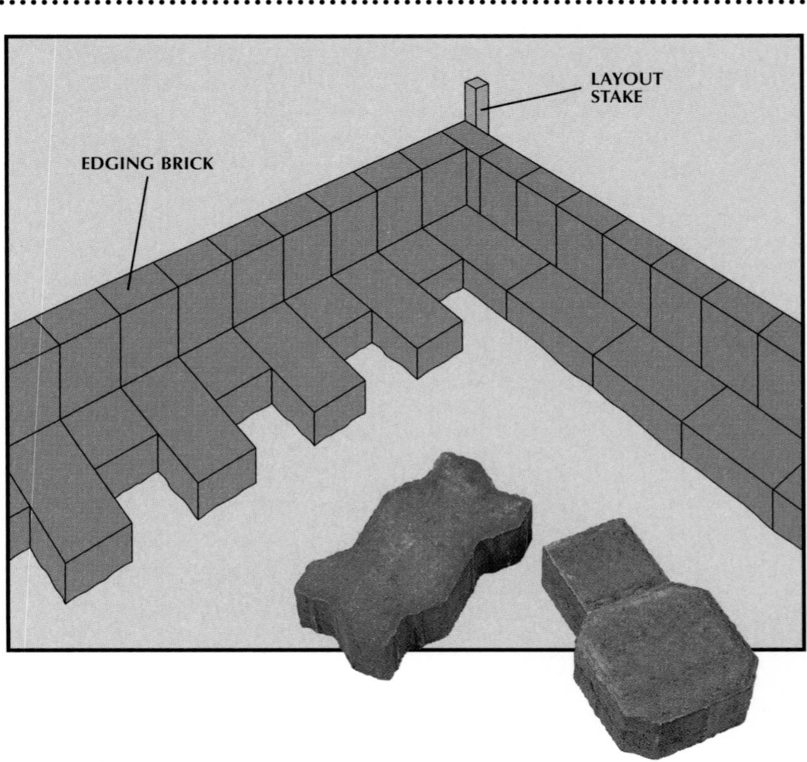

EDGING VARIATIONS

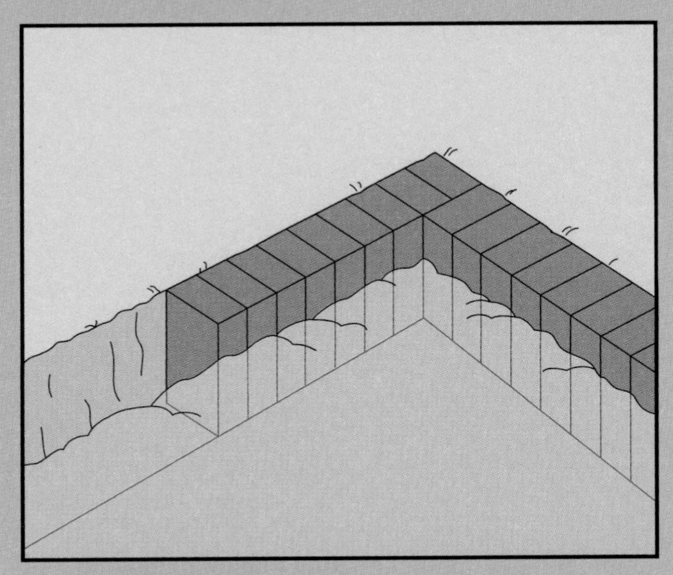

A line of soldiers.
The simplest of all brick edgings is a straight sailor course *(page 71)*; however, a line of soldiers *(above)* produces a sharper and more attractive contrast with the paving bricks—but uses almost twice as many bricks.

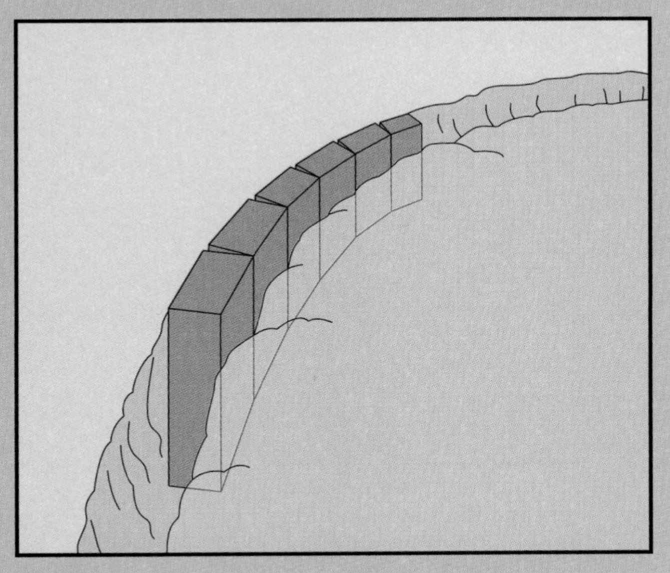

A gentle curve.
Curved brick for a curved edging is available, but expensive. You can get the same effect with rectangular bricks by angling sailors to form a gentle curve; then, fill the wedge-shaped gaps between the bricks with soil.

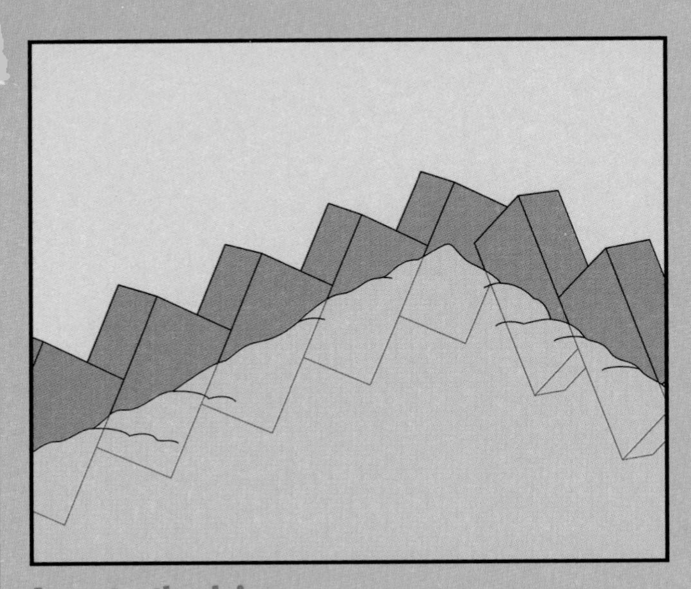

A sawtooth edging.
Half-buried sailors tilted at an angle of 45 degrees create the illusion that the edging is a row of triangular bricks neatly cut to size. The brick bases must be supported by packed earth and the tops leveled.

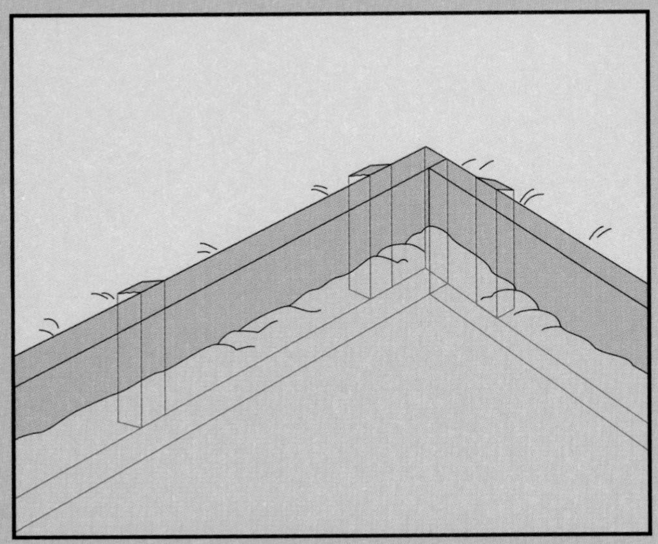

A wood edging.
The forms used for pouring concrete *(pages 43-44)* can be readily adapted for edging. Conceal pressure-treated pine or plywood edgings by setting their tops at or slightly below grade and cutting off stakes at an angle. Show off an attractive wood, such as redwood or cedar, by letting $\frac{1}{4}$ inch or so of the edging and paving bricks project above grade.

LAYING BRICKS IN SAND

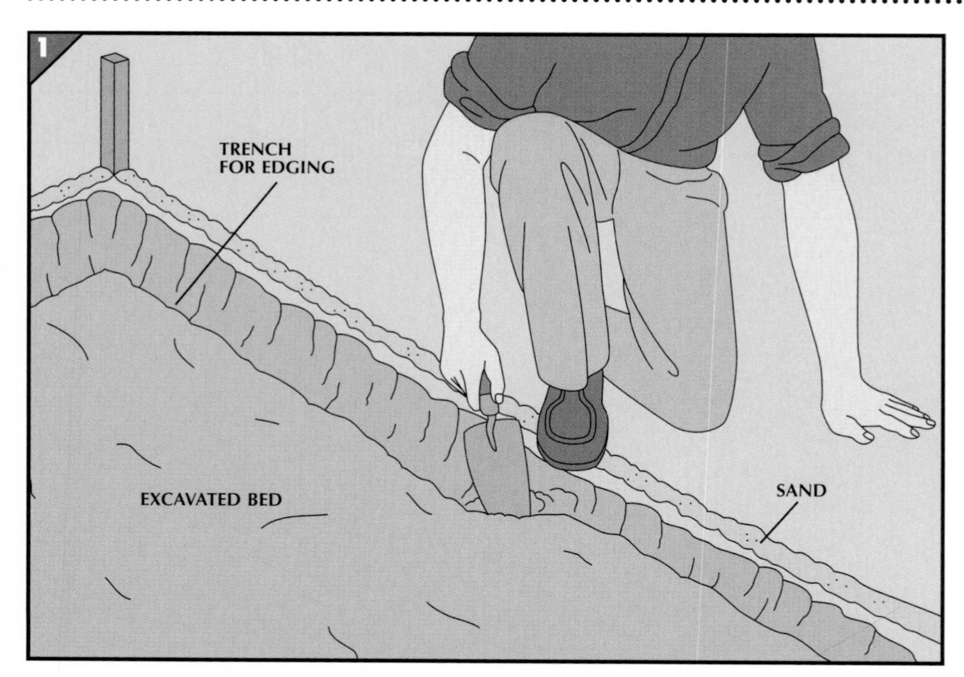

1. Edging the bed.

◆ Remove the dry-run bricks and dribble sand over the layout strings *(page 39, Step 1)* to outline the paving bed.
◆ Dig a bed $3\frac{1}{2}$ inches deep, keeping the sides of the bed as vertical as possible.
◆ For an edging of sailors, use a garden trowel to dig a trench $2\frac{2}{3}$ inches wide and an additional $4\frac{1}{2}$ inches deep along the sides of the bed *(left)*.
◆ Set edging bricks in the trench so they enclose the paving bed with their tops level with the grade. Tamp loose soil up against the bricks to hold them upright.

2. Preparing the bed.

◆ Tamp the paving bed *(page 40, Step 4)*.
◆ For a patio, set two 1-inch-thick wood strips 3 or 4 feet apart along the length of the bed.
◆ Pour concrete sand onto the bed between the strips.
◆ Smooth the sand by working a 2-by-4 screed across the strips *(above)*.
◆ Tamp the sand and screed it again if necessary, then reposition the strips and screed the next section of the bed.
◆ To prepare a bed for a path, use the method for screeding gravel for a concrete sidewalk *(page 46)*, leveling the bed so that the bricks will be flush with the edging.

3. Laying the bricks.

◆ Work the first two paving bricks into a corner of the bed.

◆ To align the courses, wrap a length of string around two bricks; then, position the bricks outside the bed so the string lines up with the inside of the brick forming the first course.

◆ Complete the course, butting the bricks together. With a level, keep the tops of the bricks at the same height; tap the bricks lightly with a trowel handle to level them, adding or removing sand from under individual bricks, if necessary.

◆ Lay subsequent courses *(above)*, repositioning the alignment bricks to line up the rows.

4. Filling the cracks.

◆ Buy masonry sand and pour a bucket of it onto the bricks.

◆ Spread the sand evenly across the surface by hand or with a brush or broom, filling the gaps between the bricks *(left)*.

◆ Gently sweep excess sand off the surface, working at a diagonal to the rows.

◆ Repeat the process, if necessary, so all gaps are filled.

MORTARING BRICKS ON A CONCRETE SLAB

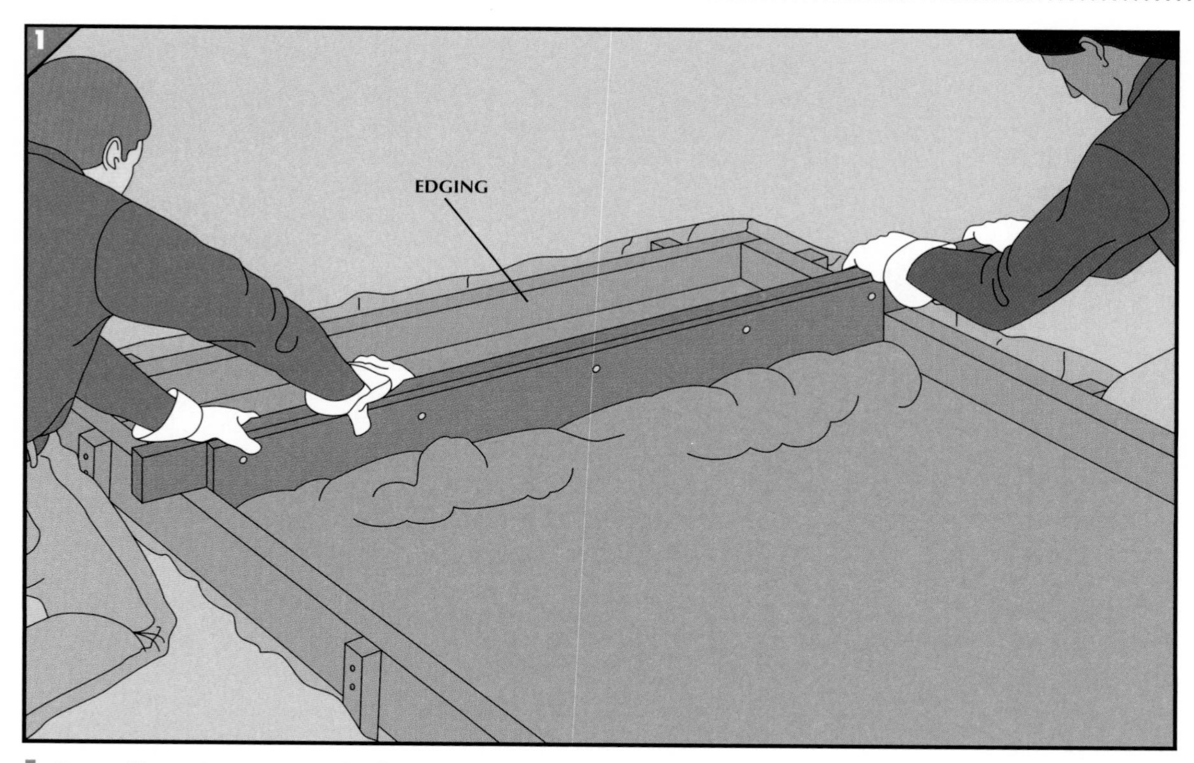

1. Screeding the mortar bed.

◆ Build temporary wood edgings staked against the outside of the slab as you would build forms for concrete *(page 43)*. The edgings should extend above the slab by the thickness of the bricks plus $\frac{1}{2}$ inch for the mortar bed.

◆ Mix a 2-cubic-foot batch of mortar *(page 10)*—enough to cover 50 square feet of slab. Shovel the mortar onto the slab.

◆ Using a screed *(page 46)* with a blade extending below the top of the edging by the thickness of the bricks, level the mortar *(above)*.

2. Laying the bricks.

◆ Soak the bricks with water.

◆ Position alignment bricks using the method on page 77, Step 3, so that there is a $\frac{1}{2}$-inch space between the bricks and the wood edgings, then set the paving bricks, smooth face up, on the mortar bed. Push each brick into the mortar and tap it lightly with a trowel handle *(left)*. Use a wood scrap to space the bricks $\frac{1}{2}$ inch apart and a level to keep their tops flush with top of the edging.

◆ For a small area, lay one complete course across the slab before starting the next; for a large one, lay rectangular segments of about 4 by 8 feet.

3. Edging the slab.

◆ For a slab above grade, remove the wood edging to make room for edging bricks to conceal the concrete and shield the paving bricks from damage and moisture.
◆ Enlarge the trench around the slab to about $2\frac{1}{2}$ inches wide and deep enough to accommodate a 2-inch layer of mortar so the tops of the edging bricks will be flush with the paving bricks.
◆ Trowel a layer of mortar 3 to 4 inches deep along the trench.
◆ Soak the edging bricks and embed them in the mortar in the trench $\frac{1}{2}$ inch apart as sailors, flat faces out *(left)*; the excess mortar will be squeezed up around the bricks.
◆ With a level, check that the bricks are flush with the paving bricks.
◆ Tamp soil against the outside of the edging bricks to pin them to the slab.

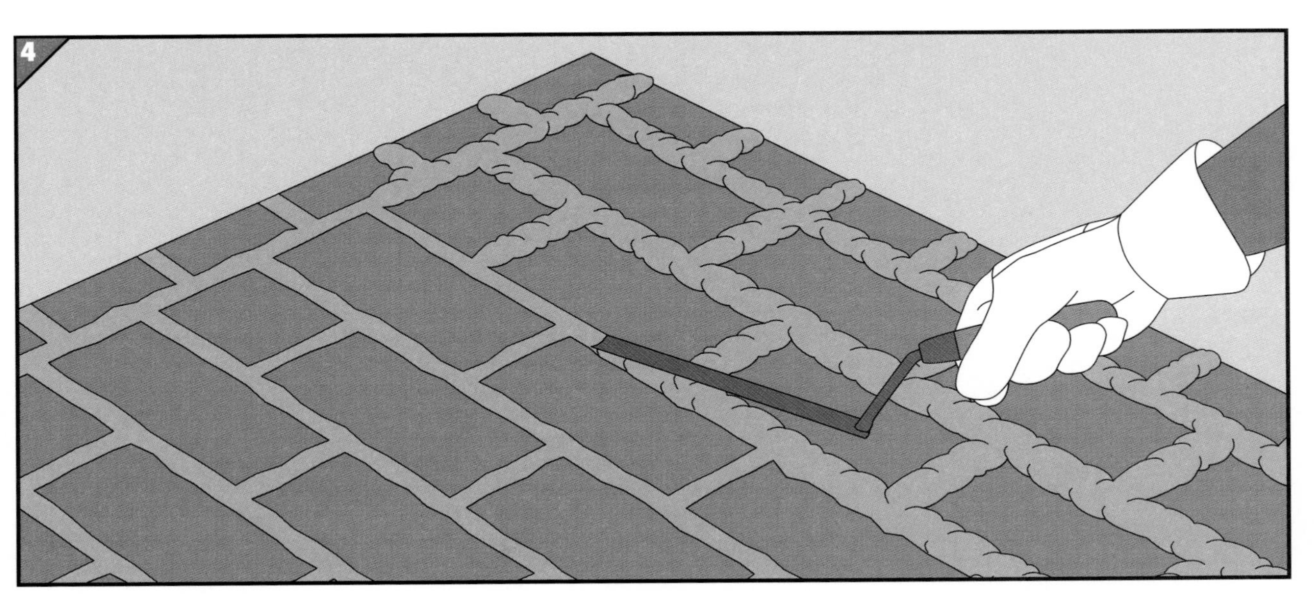

4. Grouting the joints.

◆ At least one day after laying the bricks, prepare a batch of mortar *(page 10)*.
◆ Hose down the bricks.
◆ Lay ridges of mortar on the joints and work it in with a joint filler or pointing trowel, tamping it firmly and overfilling the joints slightly.
◆ After about an hour—but before the mortar has fully hardened—remove any excess from the joints with the edge of the joint filler or trowel *(above)*. When the mortar is thumbprint hard, finish the joints with a $\frac{3}{4}$-inch convex jointer *(pages 14-15)*.
◆ After another three hours, smooth the joints with a stiff brush or a small sand-filled burlap bag. Brush mortar from the bricks and hose them down thoroughly.
◆ When the mortar has set completely —in about two days—remove any dried mortar with muriatic acid solution, adding 1 part acid to 10 parts water (or 15 parts water for light-colored brick).

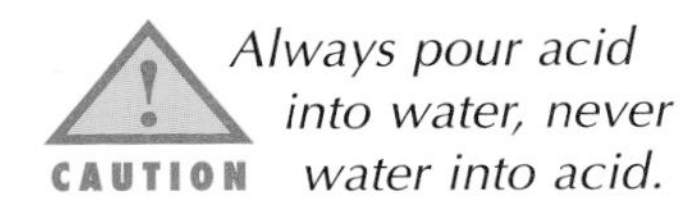

Always pour acid into water, never water into acid.

Building a Brick Wall

A freestanding wall can lend charm to a yard, enclosing spaces such as flower beds and play areas. The simplest wall to build is straight *(pages 81-87)*; a wall with corners has a few variations *(pages 88-90)*.

Choosing the Site: Start by consulting local ordinances, building codes, and your neighbors to ensure there are no legal obstacles to your plans. Next, check the soil for drainage; even a well-erected wall may buckle or sink on marshy or spongy ground. Examine the site carefully—hills and slopes present special difficulties. Avoid large trees with thick and widespread roots; and make sure that the concrete footing *(pages 63-64)*, which will be twice the width of the wall, will not encroach on an adjacent property line or sidewalk.

Building the Wall: There are several different ways of arranging the bricks, but running bond is the simplest *(page 72)*. Walls shorter than 3 feet generally do not require reinforcing *(page 82)*, but in some areas, you can build a wall up to 5 or 6 feet high without reinforcement —check local codes.

TOOLS

- Tape measure
- Maul
- Brick set
- Ball-peen hammer
- Shovel
- Hammer
- Chalk line
- Mason's trowel
- Mason's level
- Mason's line
- Line twigs and pins
- Jointer
- Carpenter's square
- Plumb bob

MATERIALS

- Lumber for story pole
- Stakes
- String
- Bricks
- 2 x 8 form boards
- Common nails (3")
- Concrete
- Mortar ingredients (Portland cement, masonry sand, lime)
- Wall ties
- Rebars
- 1 x 6s
- Sand

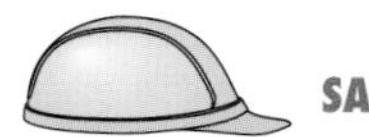

SAFETY TIPS

Mortar is caustic—wear gloves when working with it; add goggles and a dust mask when mixing it. Gloves also protect your hands from the rough edges of bricks, and hard-toed shoes prevent injury from dropped or falling brick.

A STORY POLE FOR SPACING BRICKS

To keep the courses of a brick wall on track, use a homemade measuring stick called a story pole. Cut a piece of scrap lumber to the planned height of the wall. With an indelible marker, draw a line near one end of the pole at the height of the top of the bricks in the first course; allow for a $\frac{1}{2}$-inch mortar bed plus the actual height of a brick *(page 70)*. Add a mark for each successive course to the opposite end of the pole. As you build the wall, set the pole against the newly laid bricks to make sure that the courses rise evenly.

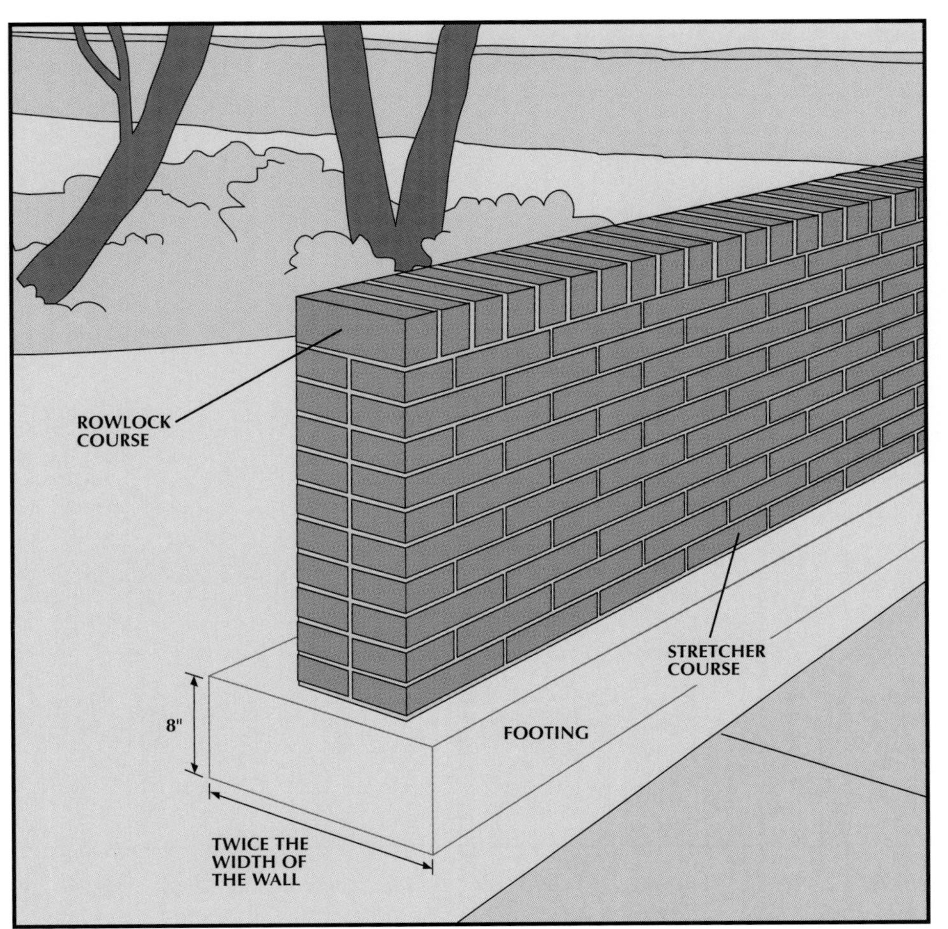

Anatomy of a brick wall.
A freestanding brick wall rests on a footing *(pages 63-64)*—a cast concrete slab 8 inches thick and twice as wide as the wall. The bottom of the footing must be 6 inches below the frost line. (The footing shown here is just below the surface as it would be in a frost-free area.) In areas with a deep frost line, you may want to build the part of the wall that is below grade from inexpensive concrete blocks.

For strength, the wall shown on these pages has two parallel layers, separated by a narrow air space, which is mortared only at the ends. The layers are bound together at regular intervals by metal strips called wall ties. The ties are placed atop every other course, starting with the second. Both layers are laid as stretcher courses, with $\frac{1}{2}$-inch-thick vertical and horizontal mortar joints. The wall is capped by a rowlock course extending from the front of the wall to the back.

PLANNING THE LAYOUT

A dry run for the first courses.

◆ From a horizontal reference line, such as the side of your house or property line, measure to the baseline you have chosen for the front of the wall.
◆ Drive stakes at the ends of this line and stretch a string between them.
◆ Lay the face course of bricks, following the string and spacing the bricks with a $\frac{1}{2}$-inch wood scrap.
◆ Place the rear, or backup, course of bricks $\frac{1}{2}$ inch behind the face course, starting with a half brick *(page 17)* and continuing with full-length stretchers.
◆ Every few bricks, set a rowlock brick crosswise *(right)*; if it is not flush with the outside edges of the front and back bricks, adjust the space between them; measure and note this distance.
◆ Finish laying the courses, ending the rear course with a half brick.
◆ Adjust the stakes to the length of the wall and measure its length.

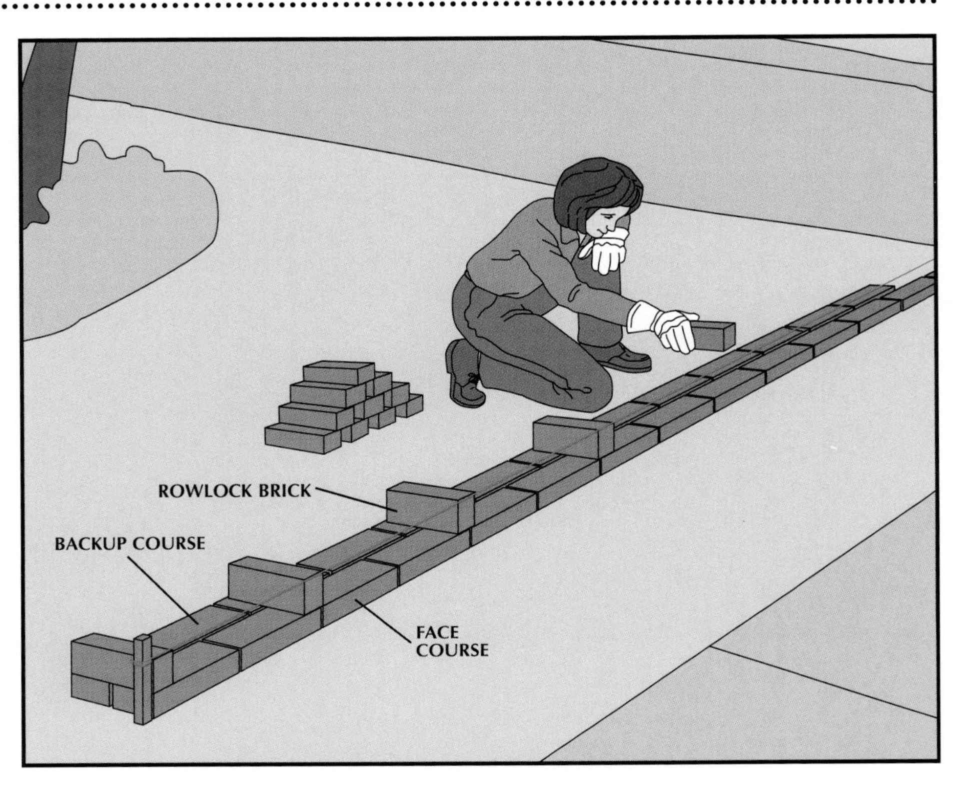

PREPARING THE FOOTING

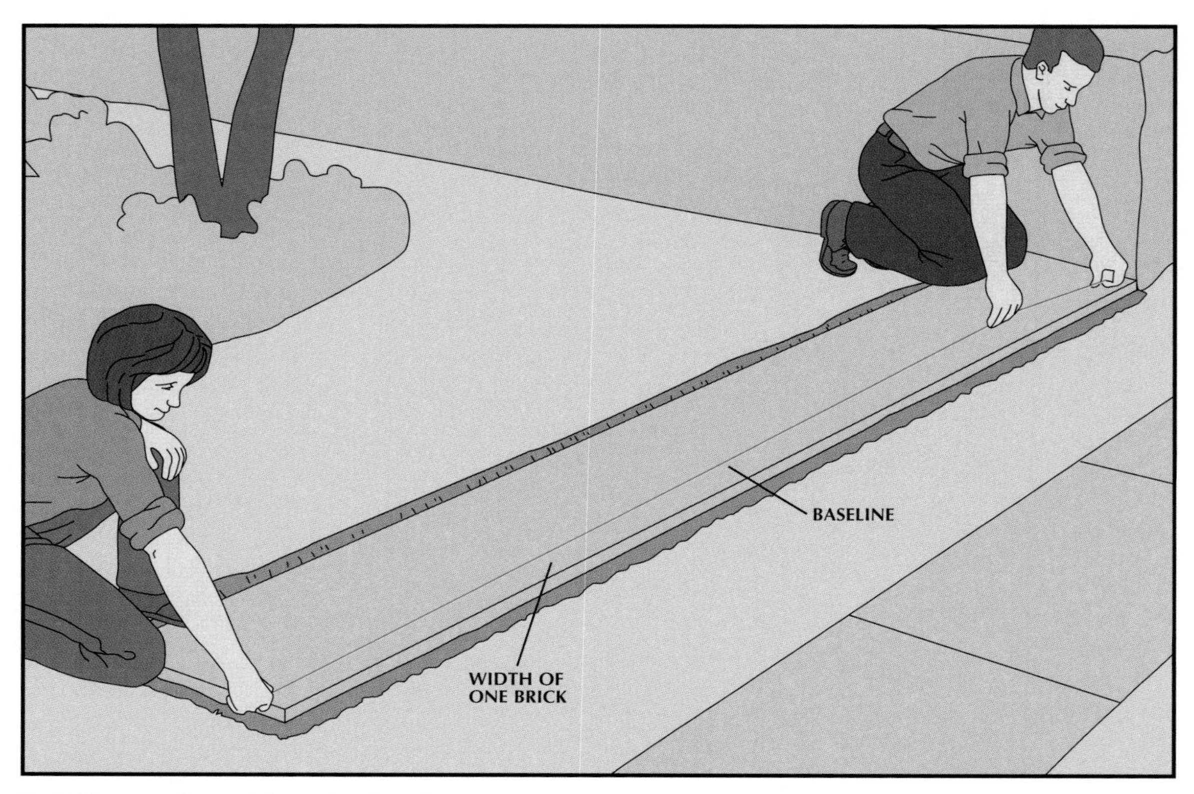

Building and marking the footing.

◆ Mark the baseline for the wall on the ground with stakes, string, and sand *(page 39)*.

◆ From this reference point, mark and dig a trench for the footing so the footing will be 8 inches (two brick widths) wider than the wall and its front edge will be offset from the front of the wall by 4 inches (one brick width). Assemble the forms and pour the concrete following the directions on pages 63 to 64. If the wall will be more than 3 to 6 feet high, add reinforcement *(below)*.

◆ With a chalk line, mark the wall's baseline on the footing *(above)*.

◆ Mark both ends of the wall on the footing, using the measurement you took after making the dry run *(page 81)*.

REINFORCING A HIGH WALL

To conform to building codes, walls over a certain height must be strengthened with steel reinforcing bars (rebars). Local codes determine the minimum height of such a wall—generally 3 to 6 feet—as well as the exact diameter and spacing of the rebars. Cast into the footing when it is poured, the first rebars extend about a foot above the slab. As the wall is built up, additional bars are tied to these with wire so the rods overlap by an amount specified by code. Once the wall is completed, the space between the face and backup courses is filled with mortar.

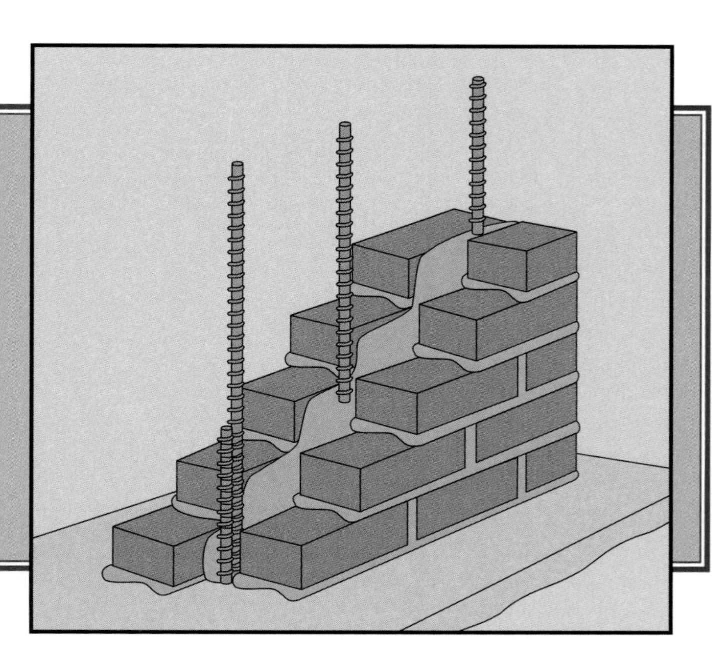

CONSTRUCTING THE LEADS

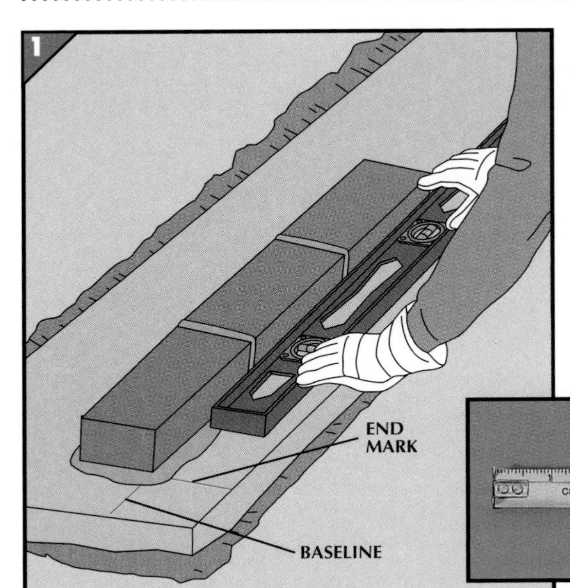

1. Laying the first bricks.

◆ Soak the bricks before laying them.
◆ Mix about 2 cubic feet of mortar *(page 10)*.
◆ Moisten 3 feet of the footing surface at one end with a fine spray of water. Let the excess water evaporate.
◆ Throw a mortar line just behind the chalk line and, starting at the end mark, lay up three bricks on the mortar bed *(pages 11-13)*. Make all mortar joints $\frac{1}{2}$ inch thick.
◆ To align the bricks with the baseline, hold a level or straightedge against the bricks *(left)* and adjust their position if necessary.
◆ Set your story pole *(page 80)* against the bricks at various points —the top of the brick should align with the first mark on the pole. Or, use a brick-spacing tape to check height *(photograph)*. Tap down on any brick that is too high; if bricks are too low, remove them and throw a new mortar bed.

2. Starting the backup course.

◆ Throw a mortar line about $\frac{1}{2}$ inch behind the front bricks.
◆ Spread a little mortar on the edge of a half brick at the outside end—this will seal the air gap between the face and backup courses at the ends of the wall. Position the brick at the end mark, spacing it from the face course by the measured amount. Continue the backup course with two stretcher bricks. (As the wall rises, half bricks will alternate between the ends of the face and backup courses.)
◆ Align the backup bricks with a level and check that the front and back bricks are level with each other.
◆ Start the second courses of face and backup bricks, beginning with a half brick for the face course and a whole brick for the backup. Lay two whole bricks on the face course and one on the backup course so there is a step up from the first to second course.
◆ Check the course heights with the story pole *(right)*.

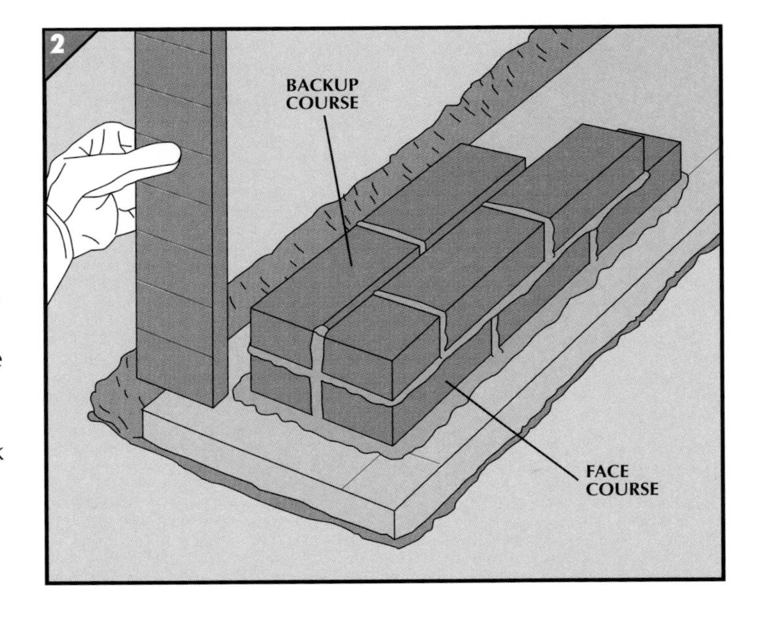

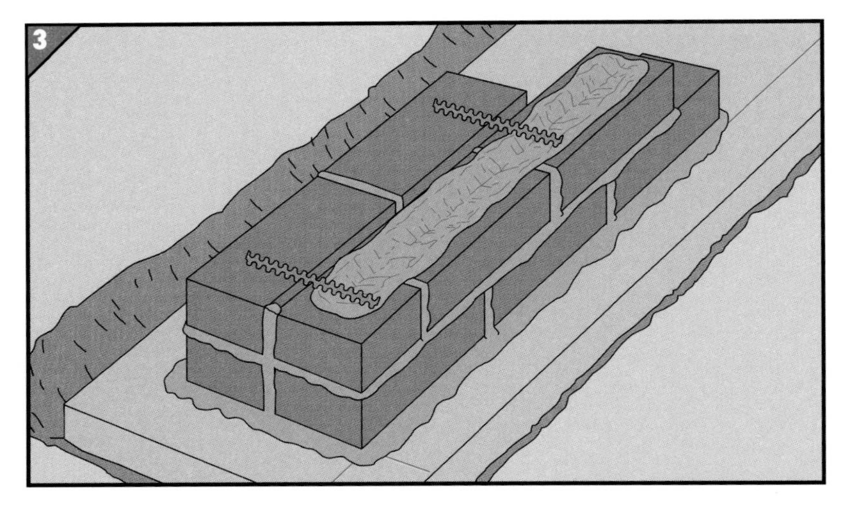

3. Placing wall ties.

◆ Throw a mortar line on the second face course and embed ties in the mortar about 12 inches apart with the free ends of the ties lying over the backup course *(left)*.
◆ Lay two whole stretcher bricks of the third face course over the ties.

4. Mortaring the ties.

◆ Once the mortar under the third face course has begun to set, bend the wall ties up away from the backup bricks, being careful not to displace any units.

◆ Throw a mortar line on the backup bricks, then bend the ties down into the mortar *(right)*.

◆ Start the third backup course—a half brick followed by a whole one.

5. Completing the first lead.

◆ Lay five face and backup courses, adding wall ties between the fourth and fifth courses. Complete the lead—the end of the wall—by placing a single brick at the end of the face course and a half brick at the end of the backup.

◆ Check with a level or straightedge for vertical alignment. Ignoring minor irregularities, gently tap any protruding brick with a trowel handle to push it into line *(left)*. Tap a recessed brick into line from behind.

6. Building the opposite lead.

At the opposite end of the footing, repeat Steps 1 through 5 to form a five-course lead. Check carefully with the story pole and the level *(right)*—unless the two leads match exactly, the wall will be unstable.

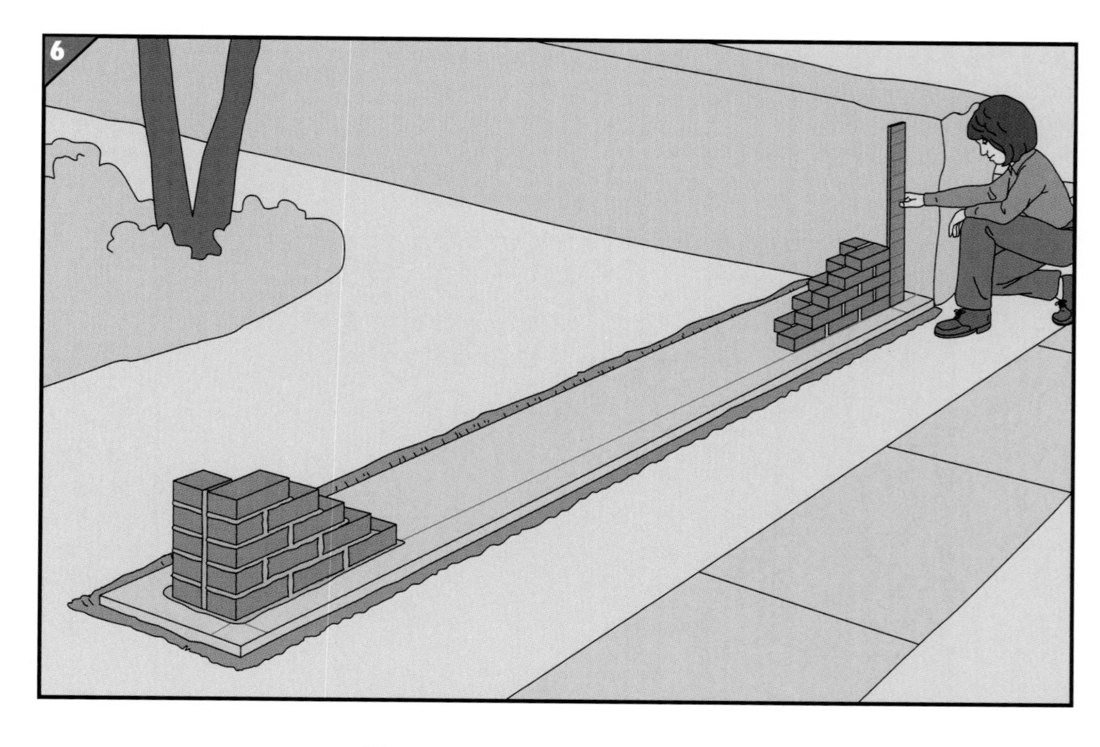

FILLING IN BETWEEN THE LEADS

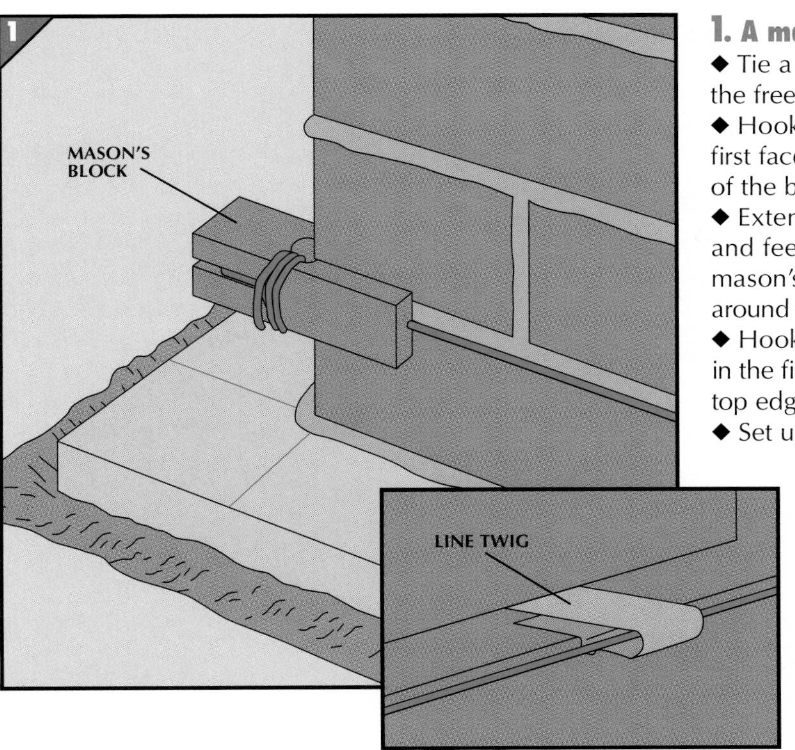

1. A mason's line for aligning courses.

◆ Tie a mason's line to a mason's block, feeding the free end of the line through the slot.
◆ Hook the block around one end brick in the first face course, aligning the string with the top of the brick.
◆ Extend the line to the other end of the wall and feed the line through the slot of another mason's block. Pull the line taut and wrap it around the block.
◆ Hook the second block around the end brick in the first course, keeping the string flush with the top edge of the brick *(left)*.
◆ Set up a second line along the rear course.

On a long wall, use a mason's tool called a line twig to support the string near the middle *(inset)*; at the center of the run, lay a brick on a piece of plywood as thick as the mortar joints, position the line twig on the brick, then set a second brick on top to hold it in place.

2. Laying bricks between the leads.

◆ Working from the ends of the wall toward the middle, complete the first face course, using the mason's line as a guide.
◆ At the center of the course, place a closure brick, buttering both ends with mortar *(page 13)*.
◆ Finish laying the first backup course the same way.

3. Building to the top of the leads.

◆ Working from the ends toward the middle of the wall, finish the next four face and backup courses *(left)*. Move the mason's line up one course at a time as you proceed and insert wall ties atop the second and fourth courses *(page 83-84, Steps 3 and 4)*.

◆ If the wall will stop at this height, add a rowlock course *(opposite)* and finish the joints *(pages 14-15)*; for a taller wall, proceed to Step 4.

4. Extending the wall upward.
Add reinforcement bars as necessary *(page 82)*; then, build new five-course leads at the ends and fill in the courses between the leads, always working from the ends toward the middle. Use a story pole as a guideline for the leads *(page 80)* and a mason's line for the bricks between them *(above)*.

CAPPING THE WALL

1. Laying the rowlocks.
◆ Starting at one end of the wall, throw mortar lines on the top face and backup course.
◆ Set the first rowlock brick at the end of the wall.
◆ Butter one side of the next brick, and lay it alongside the first with a $\frac{1}{2}$-inch joint between bricks.
◆ Lay rowlocks to the end of the wall. With standard-size bricks, every third joint in the rowlocks should align with a joint in the course below *(above)*. If not, stop laying rowlocks and lay a dry run from that point on as shown below before finishing the cap.
◆ Finish the joints in the rowlock course *(pages 14-15)*.

2. A dry run.
◆ Before throwing any more mortar, lay rowlocks to the end of the wall. Adjust the joint spacing so that the last brick rests flush with the end of the wall *(left)*.
◆ With a pencil, mark the brick locations on the wall.
◆ Remove the bricks and complete the rowlock course as described above, aligning each brick with its marks on the wall.

BUILDING A WALL WITH CORNERS

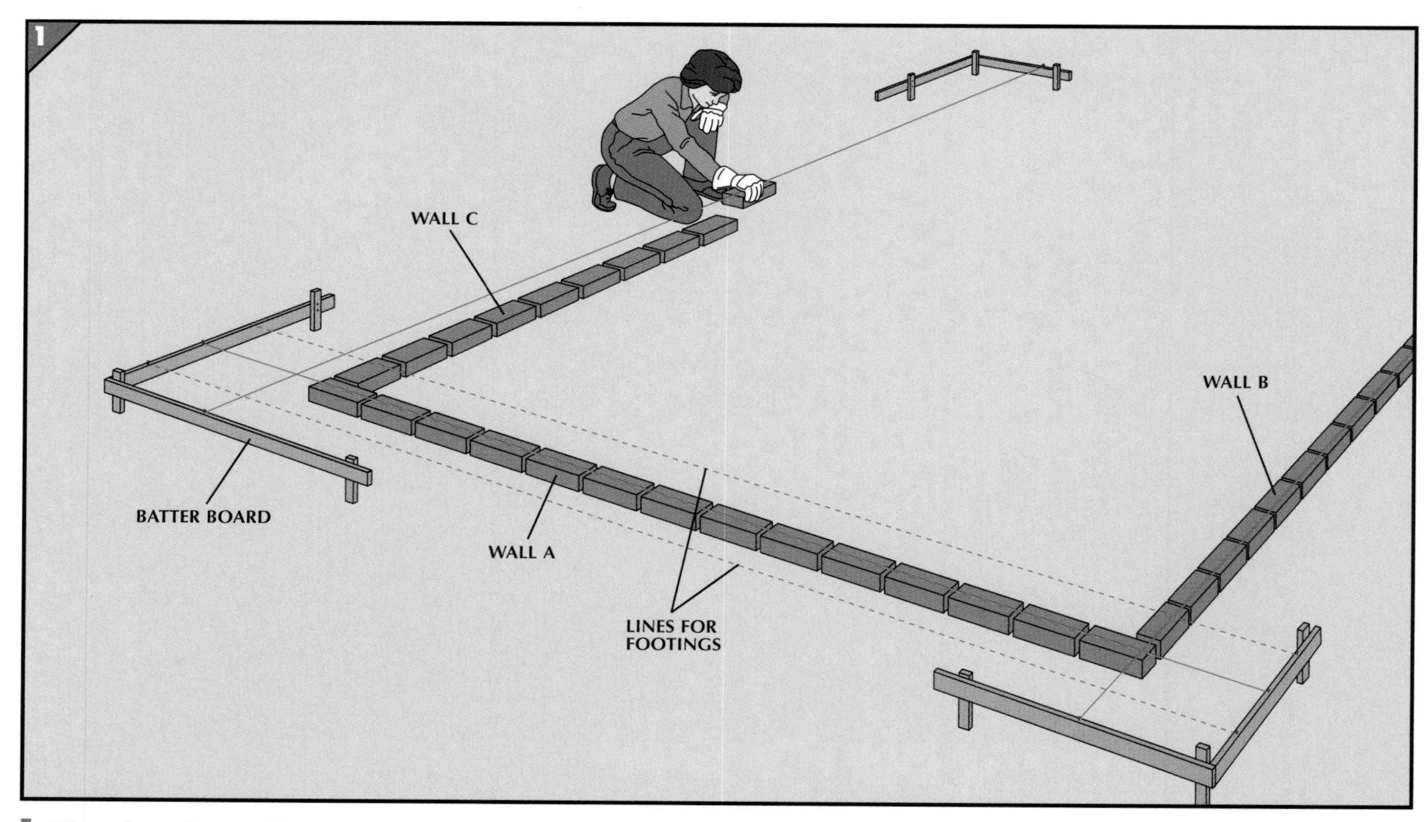

1. Planning the wall.

◆ Outside the corners of the proposed wall, set up four L-shaped sets of boards—called batter boards—each built from two 1-by-6s or 1-by-8s nailed to three stakes.

◆ Set up a string guide for the most prominent wall (Wall A in this case) and tie it to a nail in the batter board.

◆ Make a dry run of the face course of Wall A as for any straight wall *(page 81)*.

◆ Lay out the corners by setting single bricks at right angles to the ends of Wall A. Check for square with a carpenter's square.

◆ Set up string guides for Walls B and C. With a plumb bob, check that the intersection of the strings is directly above the corners of the wall. Use the triangulation method *(page 55)* to make sure the strings cross at exactly 90 degrees.

◆ Lay a dry run for Walls B and C *(above)*. With a plumb bob, mark the string at the end of these sections.

◆ Add string lines to the batter boards one brick width to each side of the wall to locate footings *(dashed lines)*.

◆ Transfer the lines for the footings to the ground with sand *(page 39, Step 1)*.

◆ Remove the strings, leaving the batter boards in place. Mark the string locations.

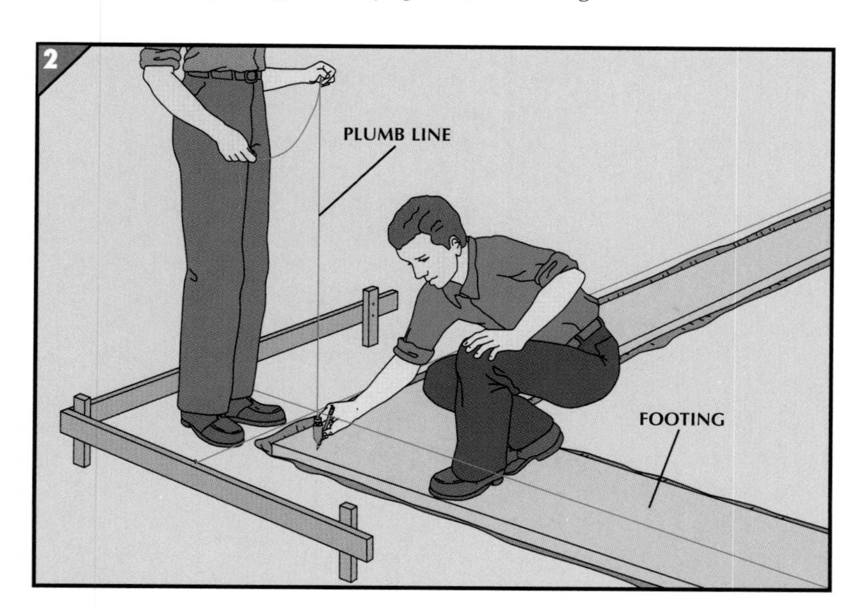

2. Marking the walls.

◆ Dig a trench, build forms, and pour concrete for each footing following the method on pages 63 and 64. If the wall will be more than 3 to 6 feet high, add reinforcement *(page 82)*.

◆ Retie the batter board strings for the front faces of the walls.

◆ Drop a plumb bob to mark the corners *(left)* and ends of the wall on the footings.

◆ On each footing, snap a chalk line between these marks to locate the face course of bricks *(page 82)*. Remove the batter boards and strings.

3. Forming the corner.

◆ Throw two mortar lines at one corner of the wall just inside the chalk lines using the technique on page 11.
◆ Lay brick A at the corner, then butter and lay brick B, using a carpenter's square to check that the bricks form a right angle.
◆ Lay the next four bricks in order—C, D, E, and F—making a six-brick corner lead.
◆ With a level or straightedge, check that the bricks are flush with the chalk lines *(right)*.

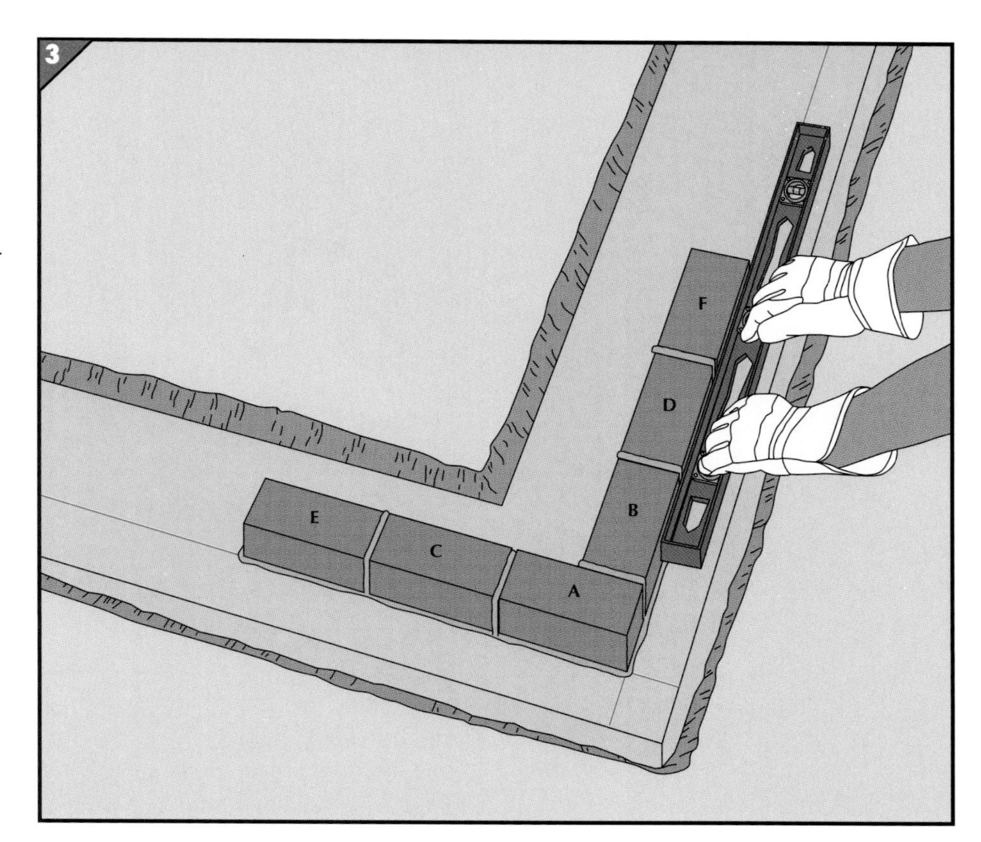

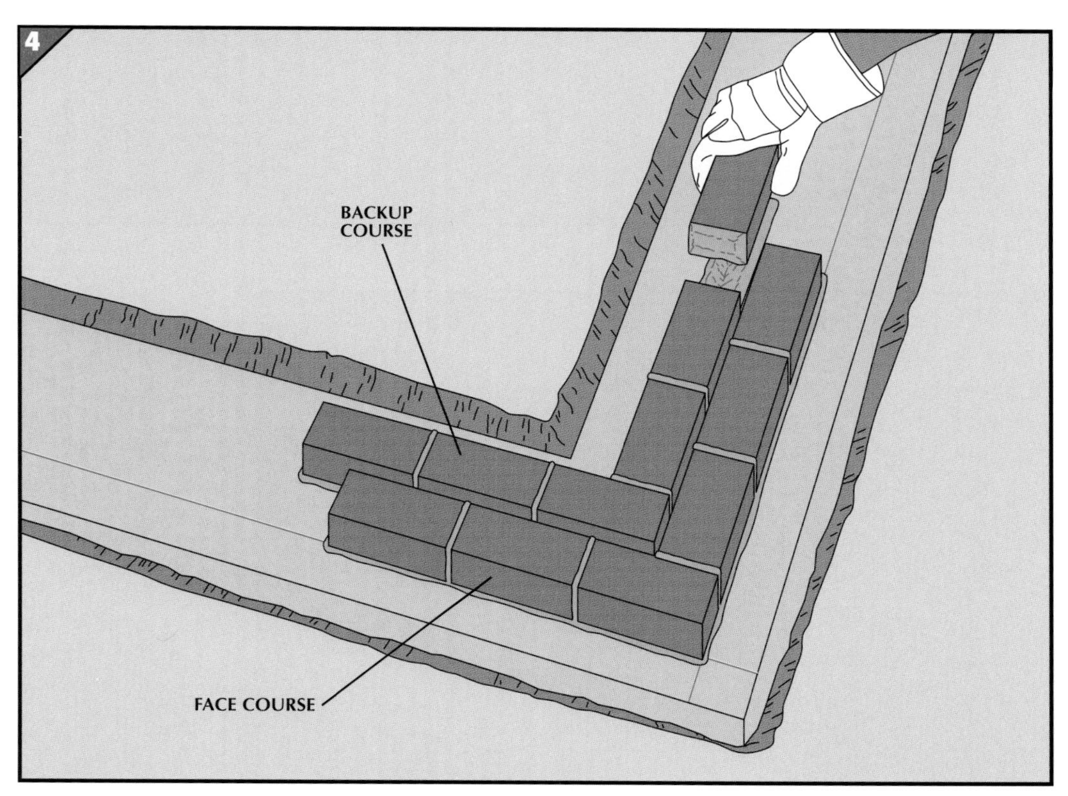

4. Starting the backup lead.

◆ Throw mortar lines behind the face course, and lay the first three bricks of the backup course spacing them $\frac{1}{2}$ inch away from the face bricks to accommodate a course of rowlocks on top of the wall, as illustrated on page 81. Check that the backup bricks form a right angle and are offset from the face bricks by a half brick *(above)*.
◆ Verify the height of the courses with the story pole *(page 80)*.

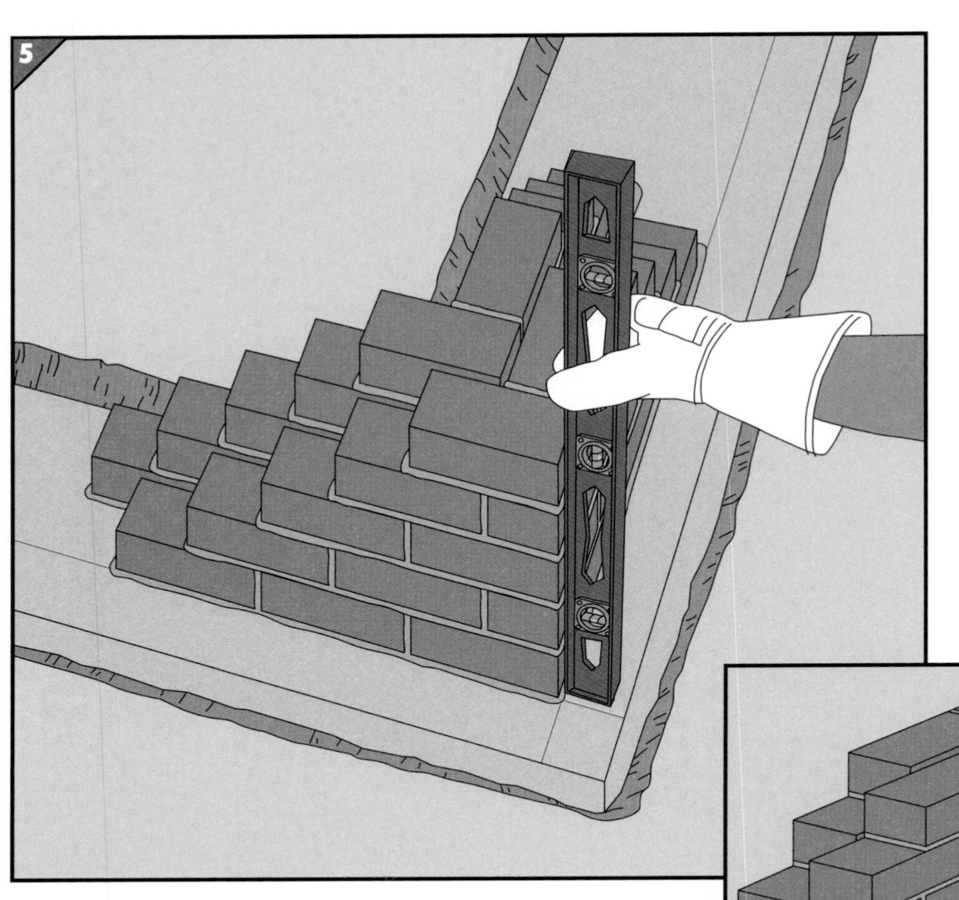

5. Completing the leads.

◆ Working on both the front and backup courses, build up the corner lead to a height of five courses, checking the alignment of the bricks *(left)*.
◆ In the same manner, build the other corner and end leads.
◆ Stretch a mason's line between the corners and fill in between the leads, adapting the techniques on pages 85 to 86. On the inside of the wall, drive line pins into mortar joints to anchor the mason's line *(inset)*.
◆ If the wall will be higher than five courses, add reinforcement *(page 82)*, then add new corner and end leads, and lay bricks between them until you reach the desired height.

LINE PIN

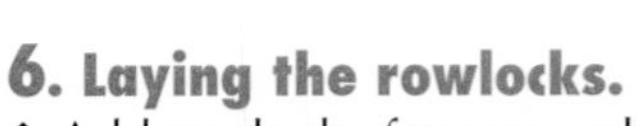

6. Laying the rowlocks.

◆ Add rowlocks for one wall as you would for a straight wall *(page 87, Steps 1 and 2)*.
◆ Lay the rowlocks for the adjoining walls so the bricks meet at the corners at right angles, as shown at right.

Veneering a Concrete Slab with Tile

An elegant way to finish a walkway or patio is with tiles. Illustrated here are ceramic tiles, but tiles made of stone, such as marble or slate, are laid in the same way.

What to Buy: Ceramic tiles suited for outdoor use are paver and quarry tile, which are 6, 8, or 12 inches square and about $\frac{3}{8}$ inch thick. The top edges of pavers are slightly rounded, while those of quarries are square edged. Buy only unglazed tiles—glazed tile will be slippery underfoot. Consult your supplier to ensure that you buy tiles suitable for your climate; or plan on sealing the tiles to prevent them from absorbing water, which can freeze in winter and crack the tiles.

To lay down the mortar bed for the tiles, you'll need a notched quarry-tile trowel—ask your tile supplier what notch size is appropriate for the tiles you buy.

Preparing the Slab: Tiles should not overlap the control joints in the slab. If you are casting a new slab *(pages 34-60)*, you can size it to accommodate an even number of tiles and to ensure that they line up with the control joints. This will prevent your having to cut the tiles to fit. An uneven base will cause the tiles to tilt or even crack; before laying tiles, check and, if necessary, smooth the surface *(page 92)*.

Making a Dry Run: To determine the exact spacing of tiles, lay out a dry run along two adjacent edges of the slab. If the tiles have lugs—built-in spacers projecting from the edges —simply place them with the lugs touching. For tiles lacking lugs, lay them $\frac{3}{8}$ inch apart using a spacer. When tiles fall short of or overlap the ends of the slab or a control joint by less than 1 inch, adjust the spacing between the tiles so they line up. If the misalignment exceeds 1 inch, lay the dry run from the middle of the slab. Mark for cutting any tiles that overlap edges or control joints.

Estimating Materials: Pavers and quarries are sold by the carton, each holding enough for 15 square feet. Figure the square footage of the slab and add 5 percent for waste. The thin-set mortar for affixing the tiles comes in 20- to 50-pound bags; 10 pounds covers 15 square feet. Grout—used to fill the joints between tiles—comes in 5- and 10-pound bags and is available in colors; one pound will cover 1 square foot. Measure the lengths of control joints in the slab and buy enough foam backer rod for filling expansion joints *(page 97)*.

TOOLS

- Rub brick
- Pointing trowel
- Quarry-tile trowel
- Broom
- Hammer
- Tile cutter
- Hacksaw frame with rod saw
- Abrasive stone
- Utility knife
- Rubber grout float
- Caulking gun

MATERIALS

- Latex-modified thin-set mortar
- Latex-modified grout
- Foam backer rod
- Self-leveling caulk

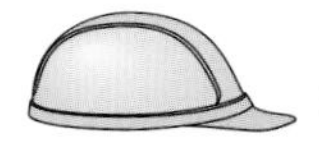

SAFETY TIPS

Mortar and grout are caustic—wear gloves when working with them; add goggles and a dust mask when mixing them. Always protect your eyes with goggles when cutting tile. Put on a dust mask when smoothing a concrete slab.

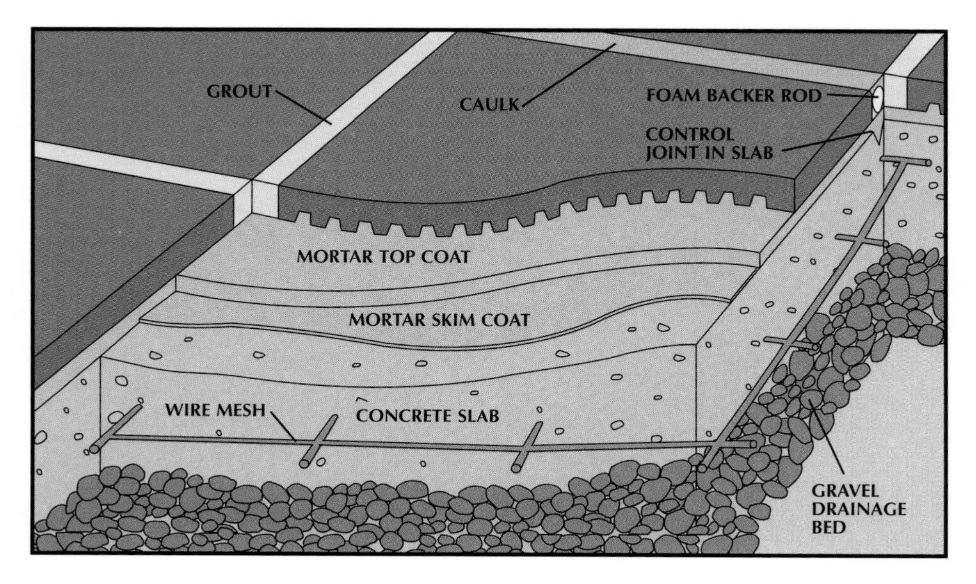

Anatomy of tile paving. Tiles are laid over a flat concrete slab reinforced with wire mesh. The slab rests on a gravel drainage bed and is sloped so water will flow off. A bed of thin-set mortar—Portland cement modified with a latex additive—comprises two layers: a skim coat that bonds the mortar to the slab, and a thicker top coat that anchors the tiles. Joints between the tiles are packed with grout; while those that line up over control joints in the slab are filled with foam backer rod and caulk, enabling tiles to move without cracking.

ESTABLISHING A SOUND BASE

Smoothing the surface.

◆ Check the flatness of the slab by slowly rolling a long piece of pipe over the surface while looking for gaps under the pipe *(above)*. With chalk, outline any low or high spots that exceed $\frac{1}{4}$ inch.

◆ Flatten high spots with a rub brick *(photograph)* or rent an electric concrete grinder.

◆ Fill in any depressions or holes, and repair cracks or spalling *(pages 19-21)*.

◆ Hose the slab clean with water.

CREATING THE MORTAR BED

1. Preparing the mortar.

◆ Pour the specified amount of liquid additive into a clean plastic tray or pail.

◆ Gradually add the thin-set mortar, stirring with a pointing trowel until the mixture is smooth *(left)*.

◆ Check the consistency by combing the mortar with the notched edge of a quarry-tile trowel—ridges should form and remain intact.

◆ Let the mortar stand for about 5 minutes and remix.

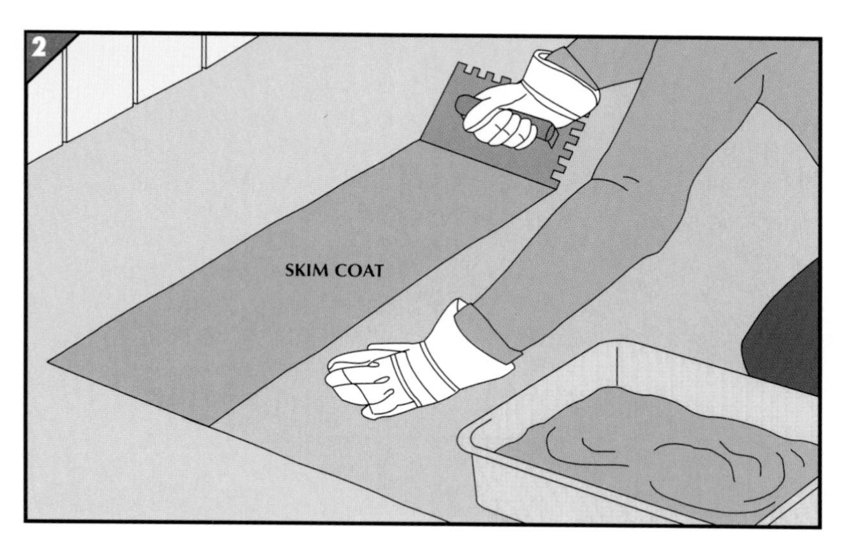

2. Applying the skim coat.

◆ Dampen the slab with a wet broom.

◆ With the flat edge of the quarry-tile trowel, scoop one-half a trowelful of mortar onto the underside of the blade.

◆ Starting at one end of the slab and holding the flat edge of the trowel against the surface at a 45-degree angle, press the tool down firmly and spread a paper-thin coating of mortar about 3 feet long *(left)*.

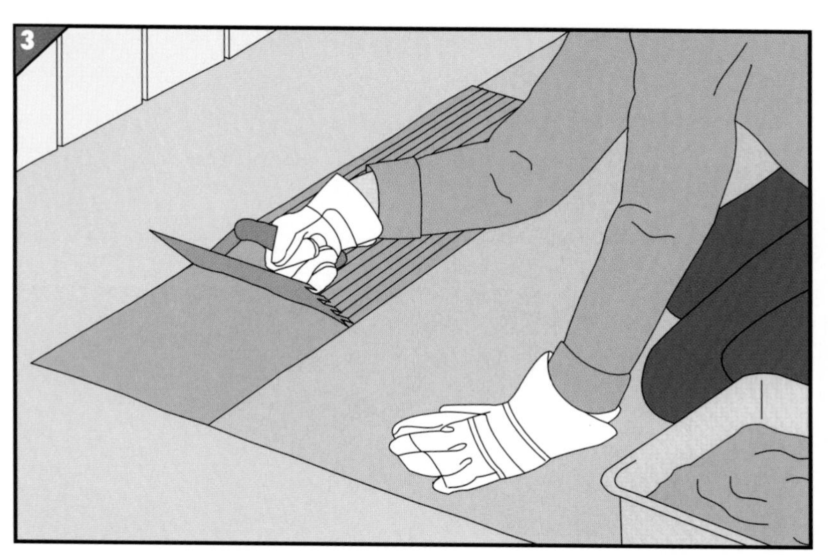

3. Applying the top coat.

◆ Before the skim coat hardens, scoop more mortar—this time a trowelful—onto the underside of the trowel, and spread a layer of mortar $\frac{1}{4}$ inch thick over the skim coat.

◆ Once the skim coat is covered, press the notched edge of the trowel against the surface at a 45-degree angle and comb the mortar, forming a series of ridges *(right)*.

LAYING THE TILE

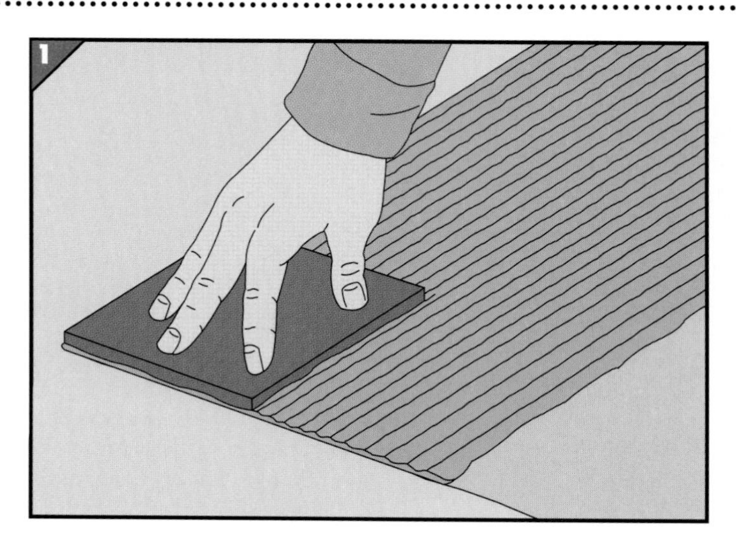

1. Seating the first tile.

◆ Position the first tile at the corner of the slab, aligning its outside edges with those of the slab.

◆ Press the tile down into the mortar bed, pushing it across the lines of mortar a little then back into position to coat its underside *(left)*.

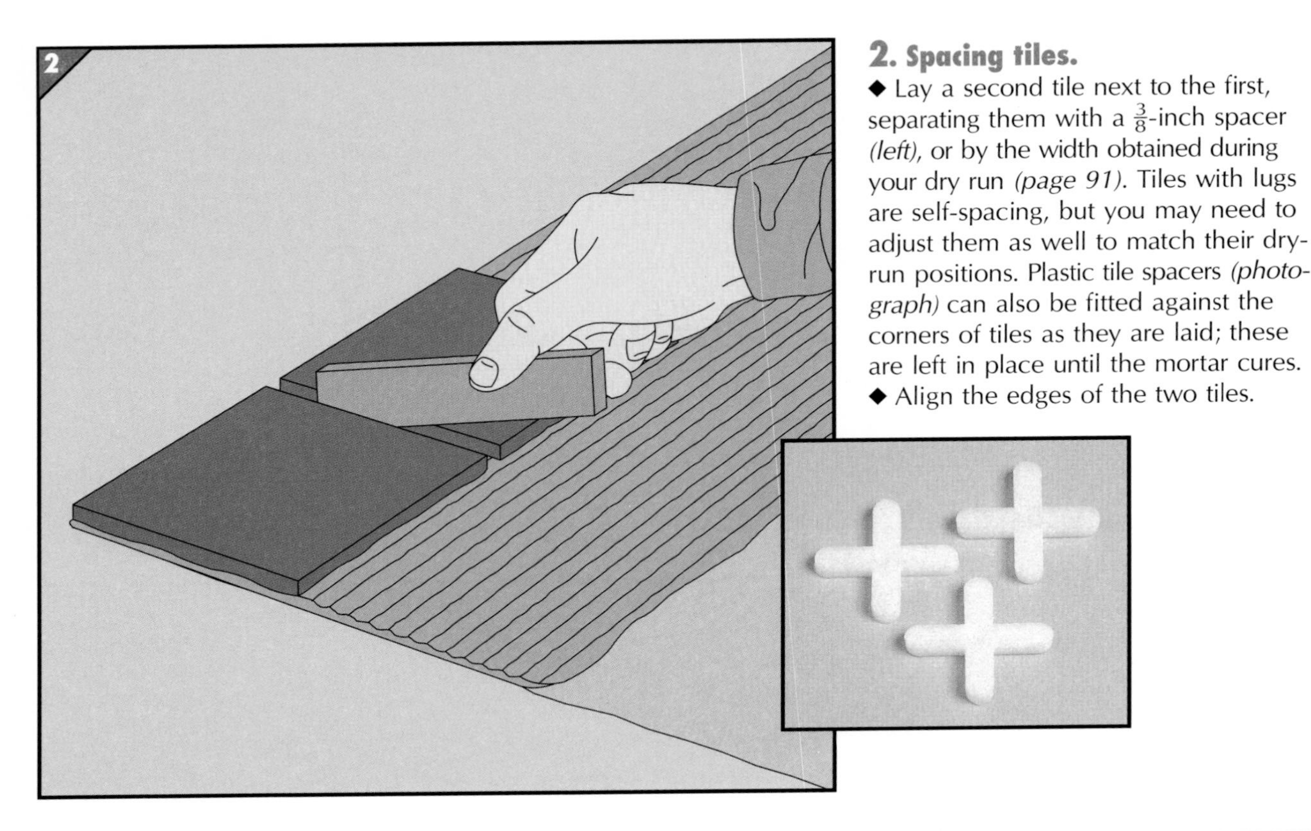

2. Spacing tiles.

◆ Lay a second tile next to the first, separating them with a $\frac{3}{8}$-inch spacer *(left)*, or by the width obtained during your dry run *(page 91)*. Tiles with lugs are self-spacing, but you may need to adjust them as well to match their dry-run positions. Plastic tile spacers *(photograph)* can also be fitted against the corners of tiles as they are laid; these are left in place until the mortar cures.
◆ Align the edges of the two tiles.

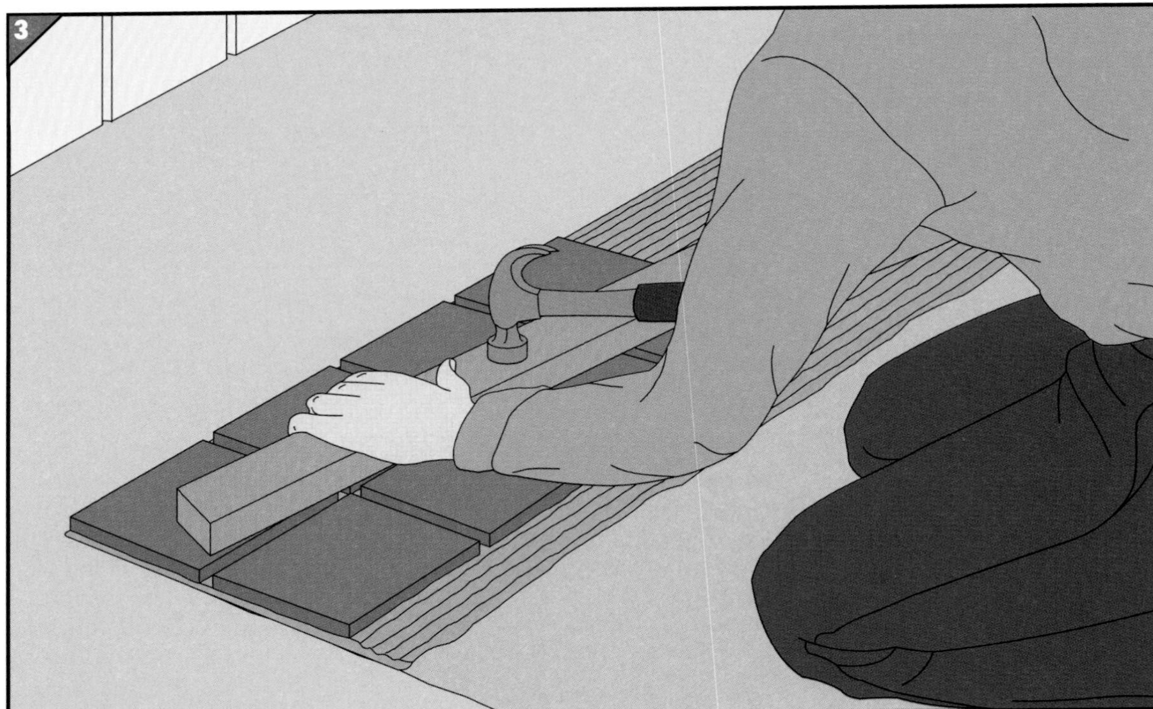

3. Truing the tile bed.

◆ After laying six or eight tiles, check that they lie flat and even by setting a long, straight board diagonally across them—there should be no gaps between the tiles and the board. To flatten the bed, tap the board gently with a hammer *(above)*.
◆ Place the board along the outside edges of the tiles and tap any crooked ones into alignment.
◆ Sponge any excess mortar from the tiles.
◆ Repeat the process across the the slab, aligning joints between tiles directly over control joints in the slab. You can stop work at any point, but be sure to clean the tiles before wrapping up. Once you have covered the slab, let the mortar cure for the amount of time specified on the package, then remove any spacers.

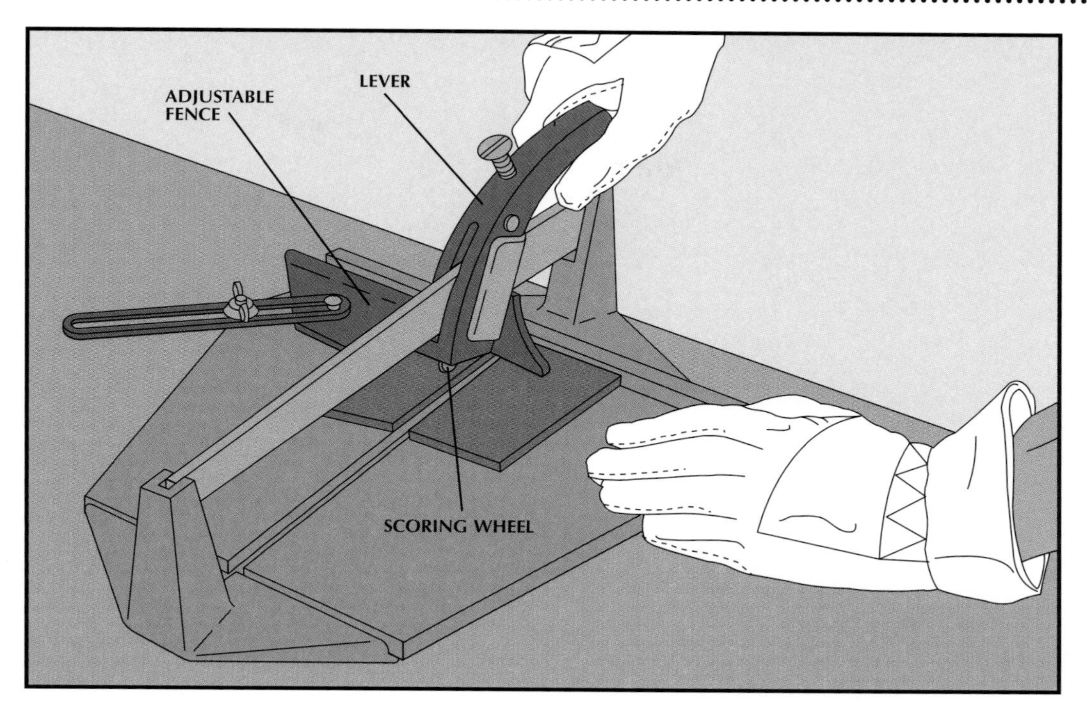

Straight cuts with a tile cutter.
◆ Set the tile on the cutter so the cutting line is directly under the scoring wheel. Adjust the fence to hold the tile in position.
◆ Draw the wheel across the tile to score it.
◆ Press down on the lever to snap the tile *(above)*.
◆ Smooth the cut edge with an abrasive stone designed for tile.

For a large job, consider renting an electric water-cooled tile saw.

A rod saw for curves.
◆ Install a rod saw in a hacksaw frame.
◆ Place the tile on the edge of a work surface and, holding the saw almost perpendicular to the workpiece, draw the blade upward a few times on the cutting line to start a kerf.
◆ Lower the saw to a 45-degree angle and cut along the line *(right)*, applying pressure on the downstroke.
◆ Smooth the cut edges with an abrasive stone designed for tile.

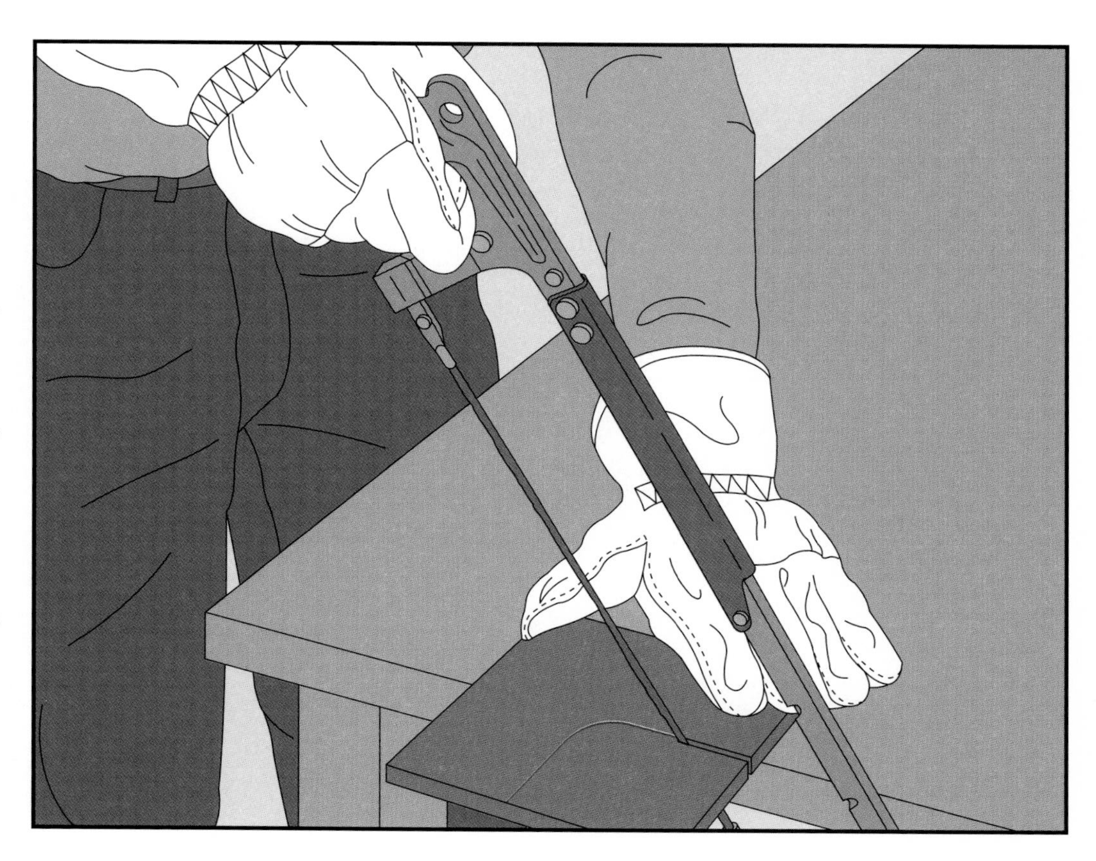

FILLING THE JOINTS

1. Preparing the grout.

◆ Into a clean pail, pour the amount of liquid latex additive recommended on the grout bag.

◆ Slowly add the grout, mixing with a pointing trowel *(left)*. Keep a record of the amounts of liquid and grout so you can duplicate them in future batches—and avoid any color variations. Continue stirring until the mixture is smooth and evenly colored.

◆ Let the grout sit for about 5 minutes, then remix it.

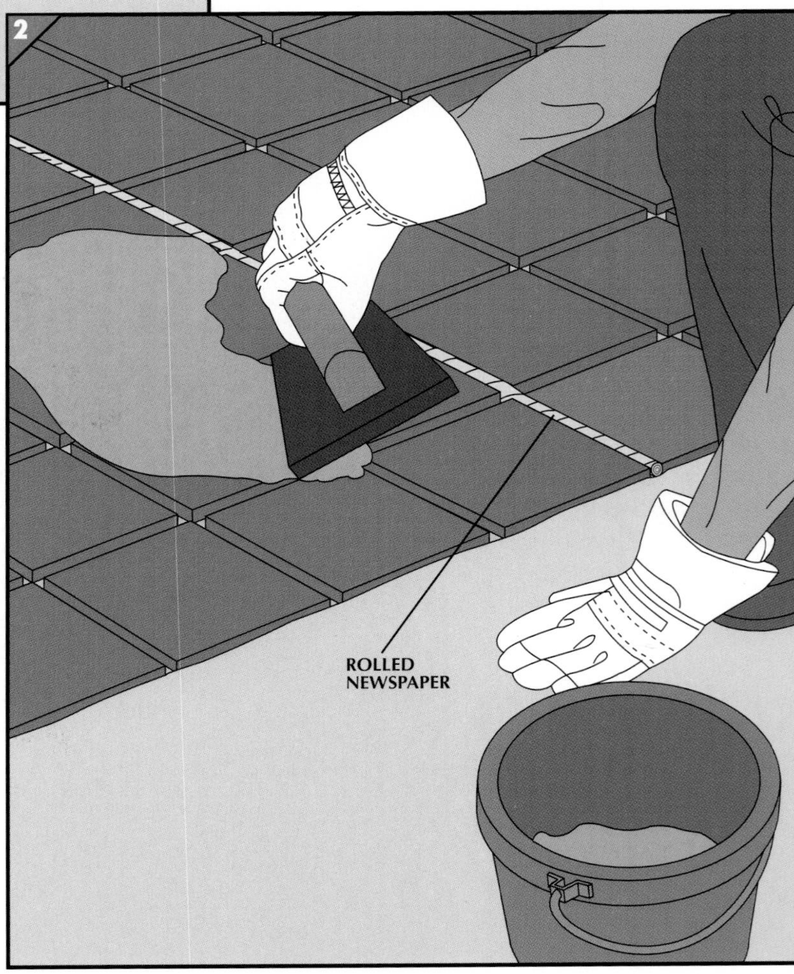

2. Grouting the joints.

◆ With a utility knife, clean mortar from the joints between the tiles.

◆ Stuff rolled newspaper into the gaps above control joints in the slab.

◆ Pour 1 or 2 cups of the grout mixture onto the tiles. Holding a rubber grout float at a 45-degree angle, drag the grout diagonally across the tiles *(right)*; press down hard enough to pack the grout into the joints and force out air pockets. Work on an area of about 5 square feet at a time.

◆ Holding the trowel almost perpendicular to the surface, scrape off excess grout. Wait 15 minutes, then wipe the surface with a damp sponge.

◆ Grout the remaining tiles the same way.

◆ Once the grout begins to harden and a haze appears on the tiles, wipe the surface clean with a soft cloth.

3. Filling the control joints.

◆ Once the grout has cured for the time specified on the bag, remove the newspaper from the control joints.

◆ Press foam backer rod slightly larger than the joints into the gaps *(left)*.

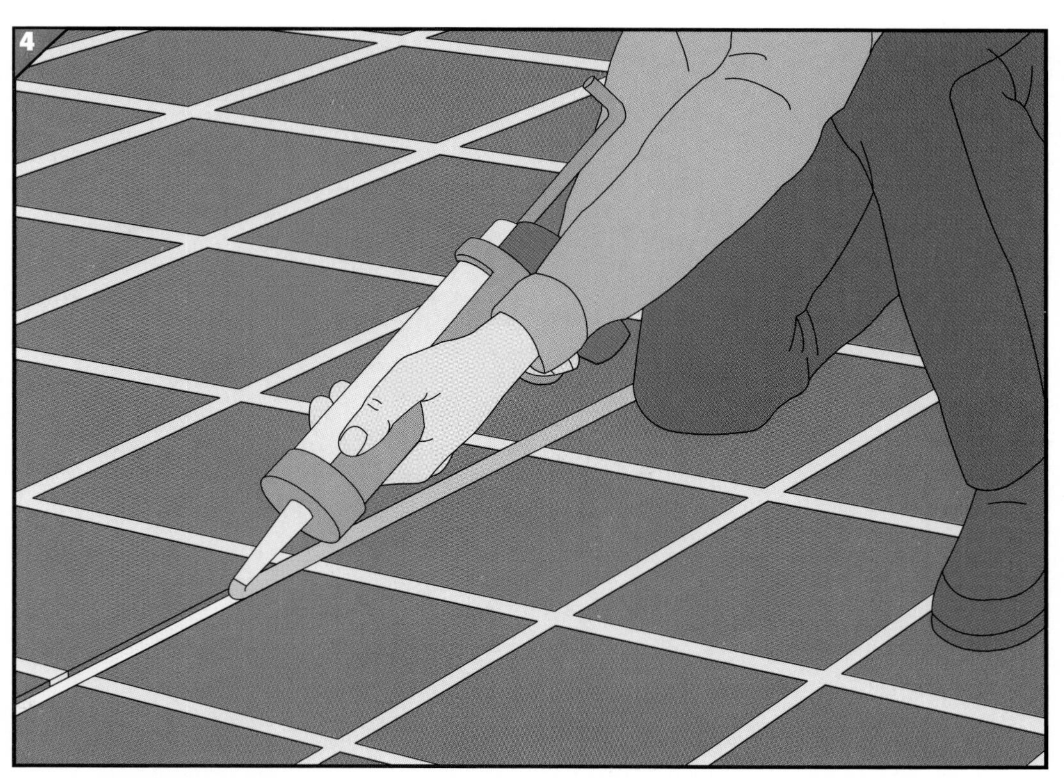

4. Caulking the joints.

◆ With a caulking gun, fill the joints with a self-leveling silicone or polyurethane caulk *(above)*.

◆ With the solvent recommended by the manufacturer, immediately wipe any caulk from the tiles.

◆ Avoid walking on the tiles until the caulk is no longer tacky.

Paving with Stone

No masonry material is more handsome or durable than flagstone, the rough-cut stone most commonly used for paving. Also known as flagging, it is available in several types *(below)*.

Paving Methods: Flagstones can be laid in sand, but the stones may shift and need to be reseated occasionally. For a more maintenance-free paving, lay the stones over a concrete slab, anchoring them in mortar and grouting the joints. Cement butter, a mixture of cement and water, is spread under each stone to strengthen its bond to the mortar. Tiles made of stone, such as marble or slate, are laid the same way as ceramic tiles *(page 91)*.

Preparing for the Job: Flagstone is sold by the square foot. Buy enough for the area to be covered, adding about 10 percent for waste. For the mortar, grout, and cement butter, estimate 150 pounds of Portland cement and 500 pounds of concrete sand—a coarser variety for mixing concrete—for every 50 square feet of paving.

Before delivery day, clear a space close to the work site and spread out tarpaulins or plastic sheeting on the lawn.

Working with Stone: Flagstones are awkward to lift—for a solid grip, wedge them up with a prybar. For moving stones, consider renting a dolly.

TOOLS

Maul
Shovel
2 x 4 screed
Tamper
Stonemason's hammer
Stone chisel
Rubber mallet
Mason's level
Stiff-bristled broom
Soft-bristled brush
Mason's trowel
Pointing trowel
Concave jointer

MATERIALS

Stakes
String
Concrete sand
Wood strips
Portland cement

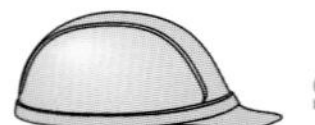

SAFETY TIPS

Wear goggles when trimming or splitting stone; you may also want to don gloves to protect your hands from the rough edges. Put on hard-toed shoes when handling stone. Mortar is caustic—wear gloves when working with it, and goggles and a dust mask when mixing it. Use goggles, a dust mask, and long sleeves and pants when working with a grinder.

Types of flagstone.
Flagging is made by splitting stone such as slate, bluestone, limestone, and sandstone into thin slabs. Your choice will be limited by the types of stone available in your area. Flagstones are commonly available in random shapes or cut roughly into rectangles as shown at right. They can also be purchased trimmed to precise patterns. Stones range in size from $\frac{1}{2}$ to 4 square feet in area and from $\frac{1}{2}$ to $2\frac{1}{2}$ inches thick. For a sand bed, flags should be $1\frac{1}{2}$ to 2 inches thick; for a mortar bed, $\frac{1}{2}$ to 1 inch thick.

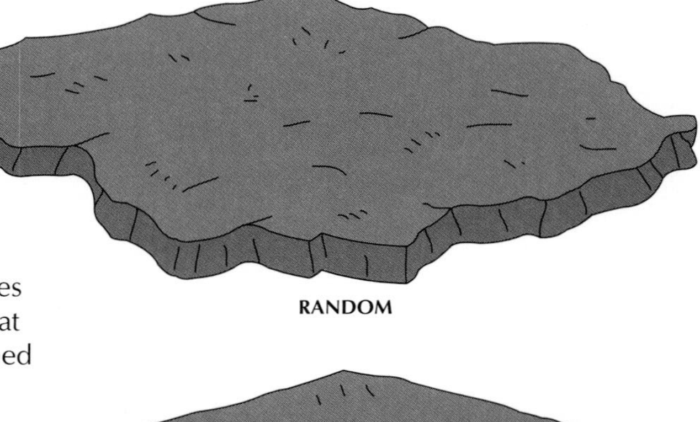

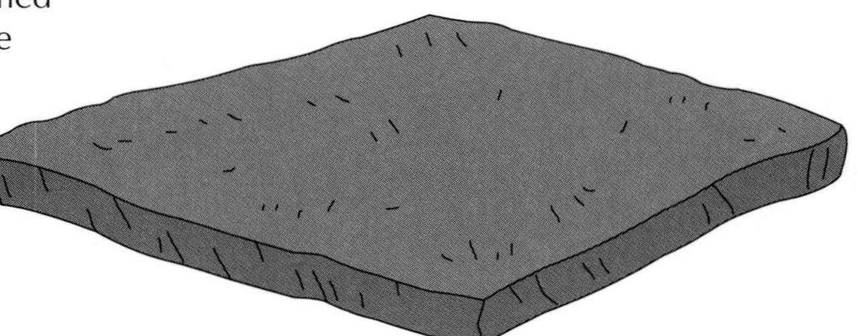

LAYING FLAGSTONES IN SAND

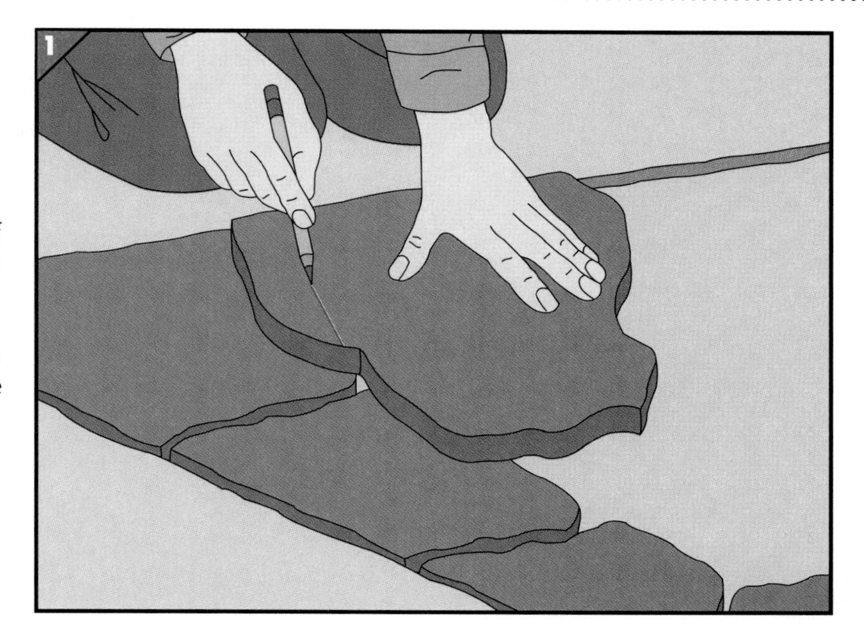

1. Fitting stones.

◆ Stake the area you plan to cover *(page 55)*, excavate it to a depth of 3 inches, and lay down a 2-inch layer of concrete sand, following the technique on page 76 *(Step 2)*.

◆ Starting at one corner, arrange six or seven flagstones on the sand bed. Line up straight edges with the edges of the bed and fit irregular edges together so joints will be about $\frac{1}{2}$ inch wide.

◆ With a pencil, mark overlapping segments to be trimmed *(right)*.

TRICKS OF THE TRADE

An Aluminum-Foil Template

To accurately mark stones for trimming, make an aluminum-foil template. Set a piece of aluminum foil in the space to be filled and cut or fold it to fit, allowing for $\frac{1}{2}$-inch joints. Set the template on a stone of approximately the right size and shape, and trace around the pattern with a pencil to mark the cutting lines.

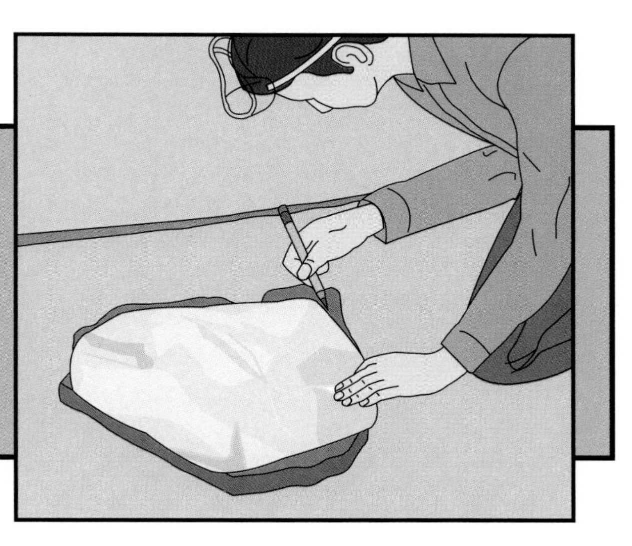

2. Trimming small segments.

◆ Place the stone on the sand bed.

◆ Chip off small pieces by striking them outside the pencil marks with a stonemason's hammer or a bricklayer's hammer *(left)*. Save the chips to serve as fillers between stones.

◆ When a segment is hard to remove, undercut it by chipping away bits from the bottom edge; then, try trimming it again.

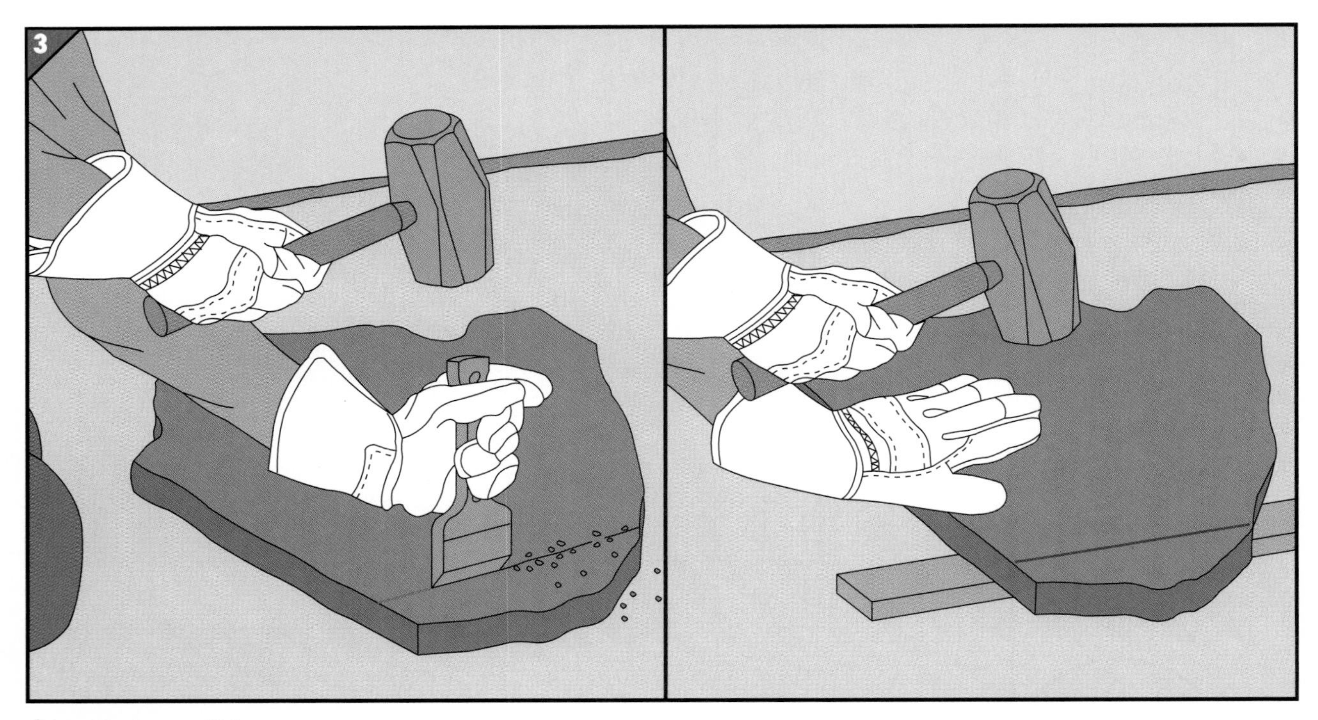

3. Splitting off large segments.

◆ Place the stone on the sand bed and score the drawn line with a maul and a stone chisel *(above, left)*; or, use a brick set in place of a chisel.

◆ Prop the stone on a board with the waste segment tilted up beyond the edge of the board. With the maul, tap the segment *(above, right)* until it falls off. If the stone does not split readily, score it along the sides and back, then prop it up and tap it again.

A RIGHT-ANGLE GRINDER FOR EASY CUTS

If cutting stone by hand seems arduous, consider renting a right-angle grinder. Designed for grinding and polishing stone, the tool can also be fitted with a diamond blade for cutting. Wear the safety gear recommended on page 98, and practice on scrap stone until you are comfortable with the technique. Turn on the grinder with the blade clear of the stone and, holding the tool upright, score the cutting line with a light pass; then apply gentle pressure to make the cut.

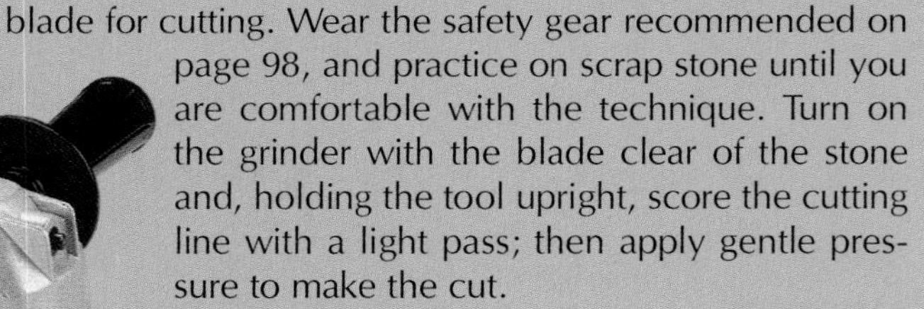

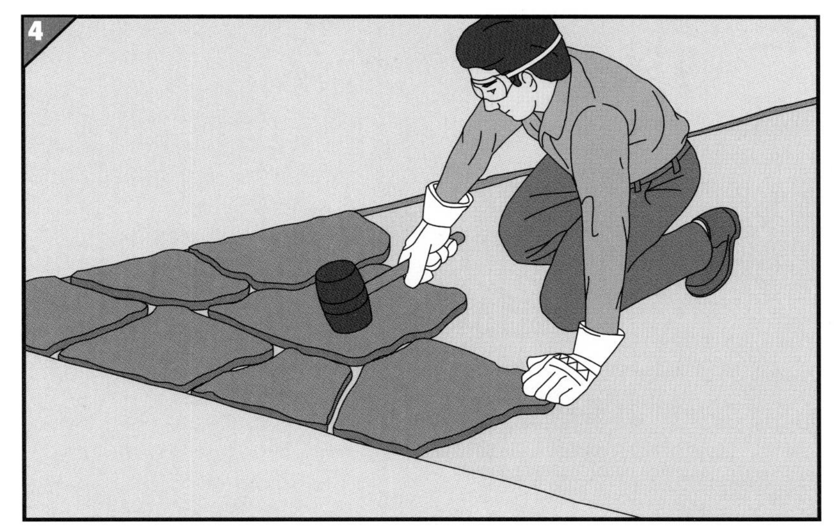

4. Embedding the stones.
With a rubber mallet, tap each stone down into the sand bed so its top lies about $\frac{1}{2}$ inch aboveground *(left)*.

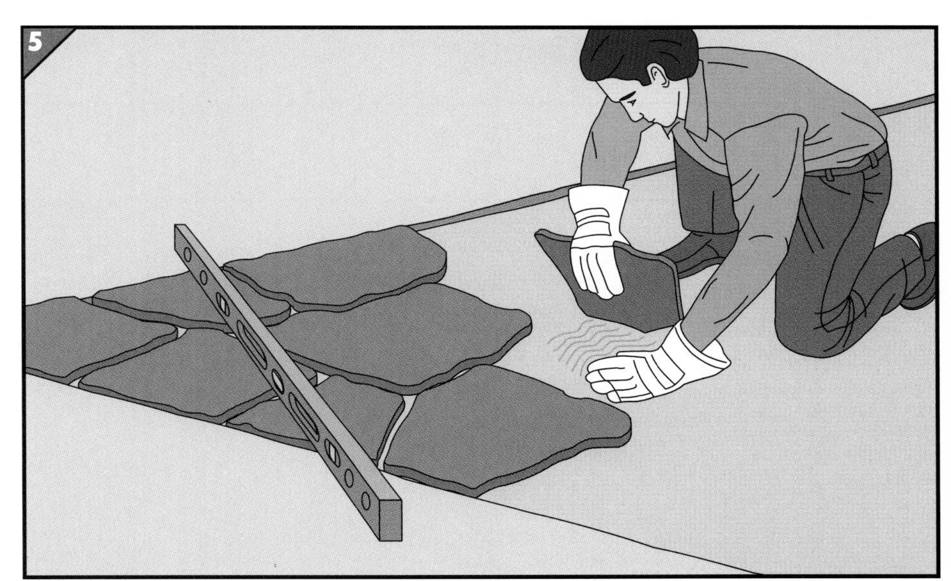

5. Truing the surface.
◆ Set a mason's level across the stones to check if they are even.
◆ Brush additional sand under low stones and scoop sand out from under high ones *(right)*.
◆ Repeat Steps 1 to 5 until you have covered the entire bed with stones.

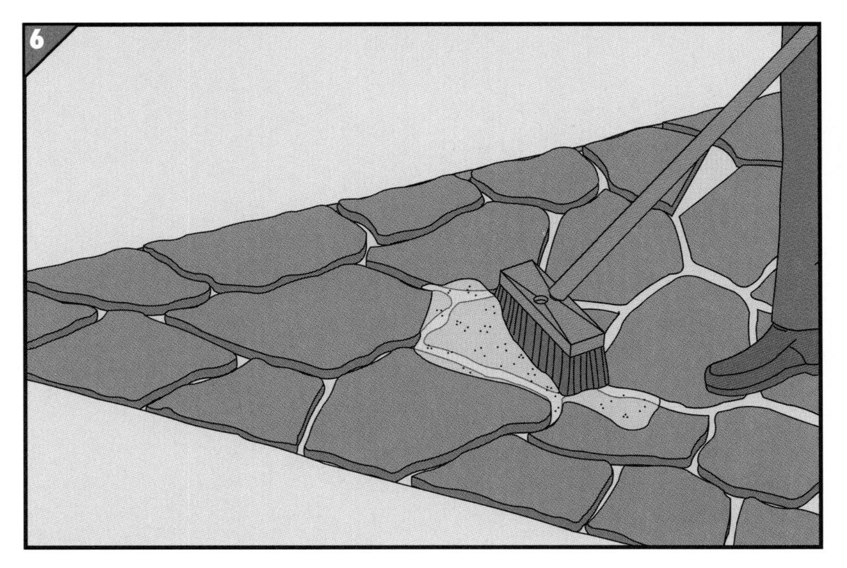

6. Filling the joints.
◆ Shovel more sand over the flagging and, with a stiff-bristled broom, sweep it across the stones to fill the joints to the top.
◆ Wet the surface with a fine water spray and let it dry.
◆ Repeat the joint-filling and spraying process until the joints are flush with the stones.

SETTING FLAGS IN A BED OF MORTAR

1. Preparing the mortar bed.

◆ If you do not have an existing concrete slab, cast one as described on pages 54 to 58.

◆ Arrange stones on the slab in a dry run, leaving no more than $\frac{3}{4}$ inch between them. Trim the stones as necessary *(pages 99-100)*.

◆ In one container, prepare a batch of mortar from one part Portland cement, four parts masonry sand, and just enough water so the mortar holds the shape of a ball when you grasp it.

◆ In a separate container, make cement butter: Portland cement blended with enough water to give it the consistency of soft butter.

◆ Remove three or four stones from the slab and wet the concrete with a soft-bristled brush *(left)*.

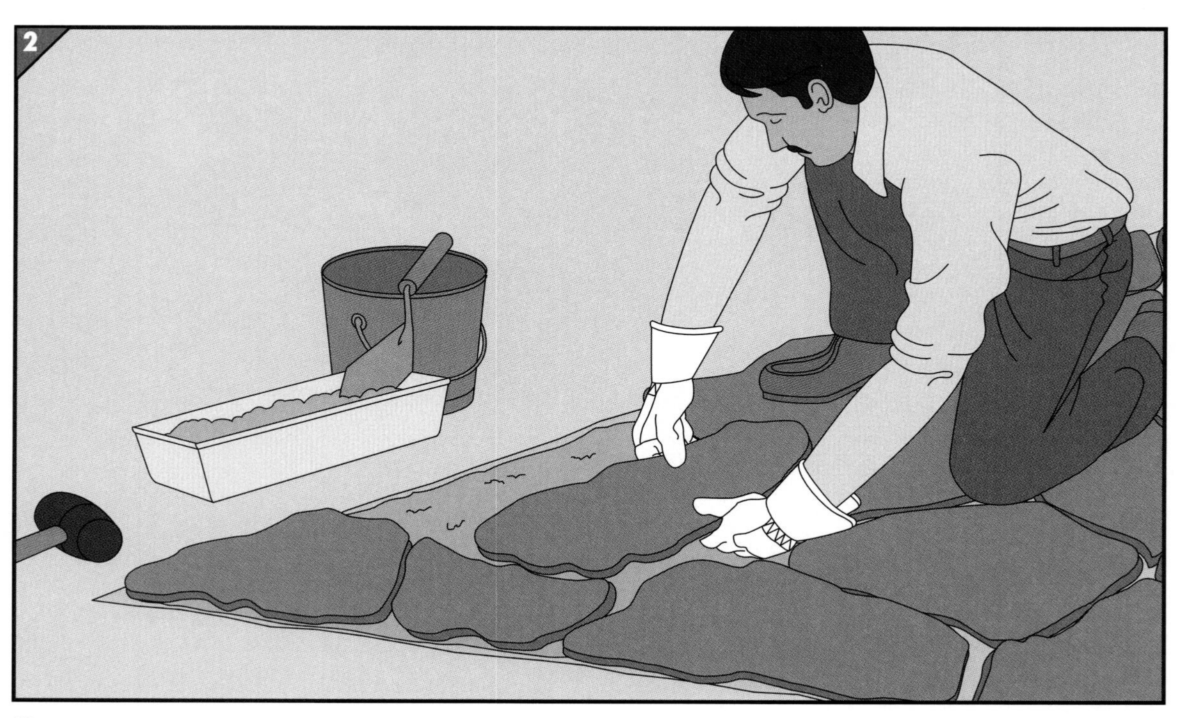

2. Setting the stones.

◆ With a mason's trowel, spread the mortar 1 inch thick over the dampened section of the slab.

◆ Reposition the stones *(above)* and, with a rubber mallet, tap them down about $\frac{1}{2}$ inch into the mortar.

◆ When you have set about a dozen stones in the mortar, level them *(page 101, Step 5)*.

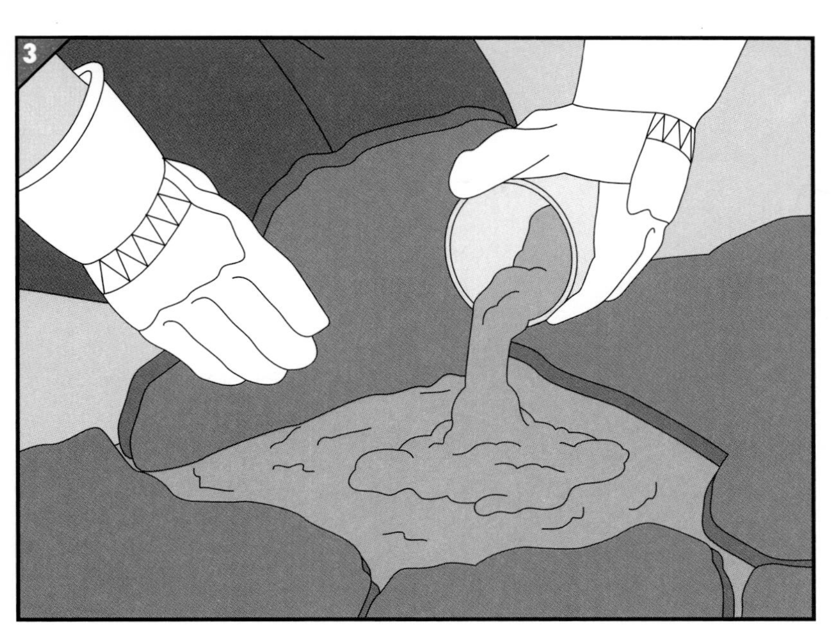

3. Applying cement butter.

◆ Immediately after leveling the stones, pick up one flagstone at a time from the mortar bed and, with a small container, dribble about $\frac{1}{4}$ cup of cement butter evenly over the bed *(left)*.

◆ Replace the stone and tap it back into position with the mallet.

◆ When you have buttered all the stones, recheck the level.

4. Raking the joints.

◆ With the tip of a pointing trowel, pack the mortar between the stones under their edges *(right)*.

◆ Scrape out excess mortar so the bottoms of the joints are about at the level of the bottoms of the stones. Sponge mortar off the flagging.

◆ Let the mortar cure for 24 hours.

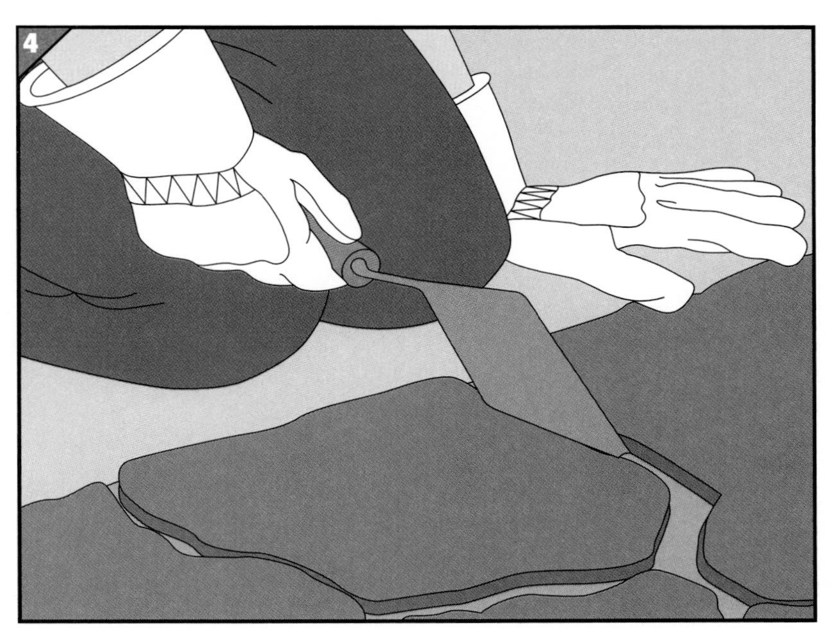

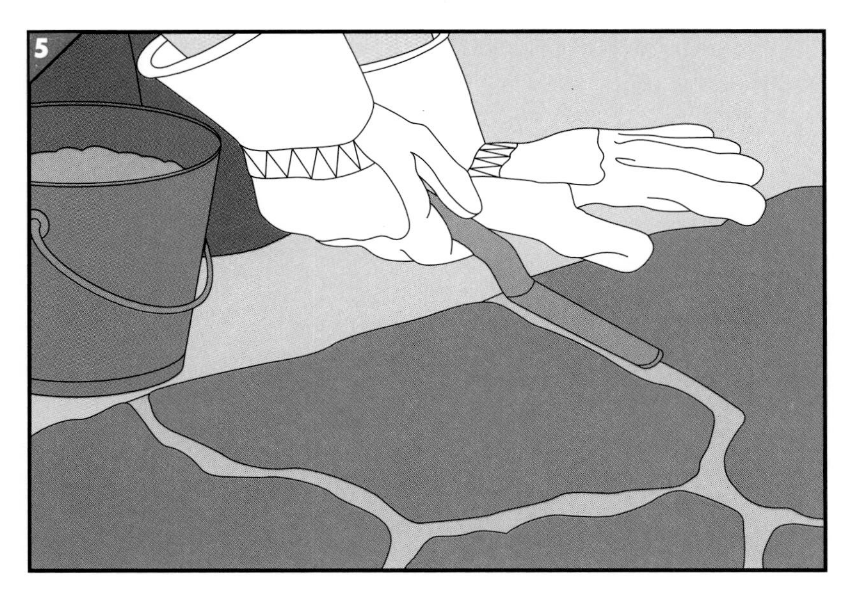

5. Mortaring the joints.

◆ Prepare a batch of mortar *(pages 10-11)*.

◆ With the tip of a concave jointer, push mortar into the joints between the flagstones and compact it *(left)*.

◆ Sponge off the excess; let the mortar dry for 24 hours before walking on the paving.

Constructing Walls of Natural Stone

Few masonry materials are more attractive or enduring than stone. Natural or quarried rubble produces the most rustic-looking walls, but requires trial and error to arrange. Square-cut stones that create gridded wall patterns need less experimentation as you go along. A simple dry wall *(pages 106-108)*—requiring neither mortar nor footing—may need occasional maintenance if stones become dislodged; a wet wall *(pages 109-111)*—laid over a footing and mortared—is relatively maintenance-free. If you plan on building a wall higher than 3 feet, check local building codes for restrictions.

A New England Dry Wall: Built directly on the ground or on a bed of gravel, a dry stone wall has overlapping joints like a running-bond brick wall. The wall slopes in two directions—the base is wider and longer than the top—and stones tilt downward toward the center so the gravitational pull compacts the wall and keeps it intact. You can use stones picked up from fields, but they may require considerable cutting to make them join securely. Most builders buy quarried rubble; easily workable types—bluestone, sandstone, or limestone—are best. Don't be too meticulous; a rough wall generally looks better and is sturdier than a fussily even one.

A Mortared Wet Wall: This style of construction is often used for retaining walls, and requires weep holes and gravel fill to drain off water. Like a dry wall, it is tapered inward to keep the stones in place. Stones that are roughly rectangular are easier to build with than rubble; mortar them together over a 6-inch-thick concrete base 18 inches wide or less.

Estimating Materials: Stone for walls is sold either by volume or weight. To calculate your needs in cubic yards, multiply the length, width, and height of the wall in feet, divide the result by 27, and add 10 percent for waste. For the amount of stone needed in pounds, multiply the volume in cubic feet by 125. For a rough estimate of mortar needed, divide the volume of the wall by 5.

TOOLS

- Shovel
- Tamper
- 1 x 2s for slope gauge
- Level
- Stonemason's hammer
- Mason's trowel

MATERIALS

- Gravel
- Plastic tubing (1")
- Mortar ingredients (Portland cement, masonry sand, hydrated lime)

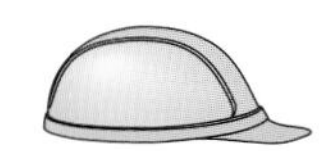

SAFETY TIPS

Wear gloves and hard-toed shoes when handling stones. Put on gloves to work with mortar, and goggles and a dust mask when mixing it.

Rubble vs. square-cut stone.
Rubble is uncut stone, either quarried or obtained as natural field and river stones. The rough surface of quarried rubble holds mortar better than the worn surfaces of field or river stones. Most dealers sell only quarried rubble, in pieces 6 to 18 inches in diameter.

Square-cut stone is roughly trimmed. A more precisely shaped—and expensive—version is called ashlar. Widths generally range from 3 to 5 inches, heights from 2 to 8 inches, and lengths from 1 to 4 feet.

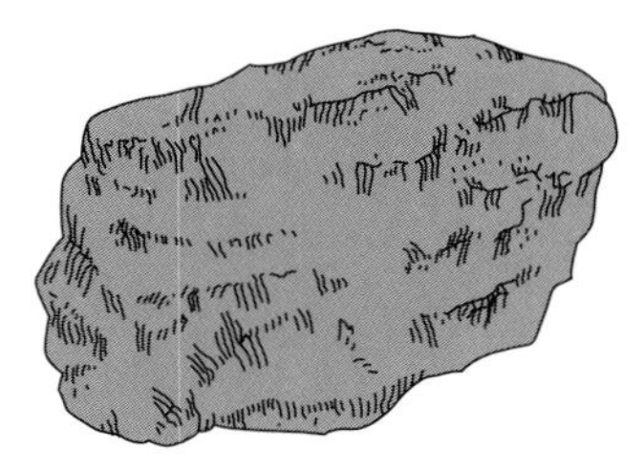

QUARRIED RUBBLE

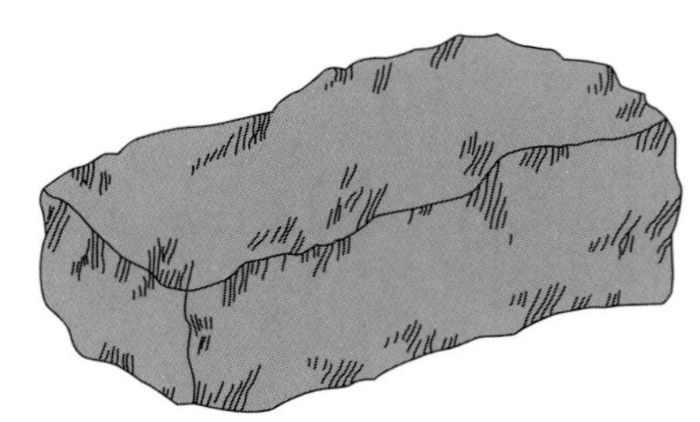

ROUGH SQUARE-CUT

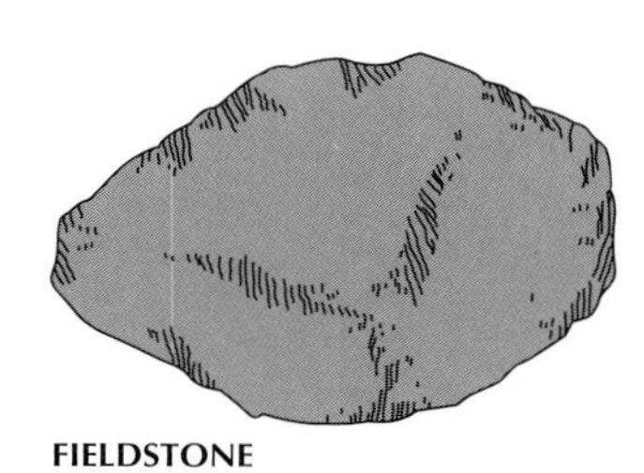

FIELDSTONE

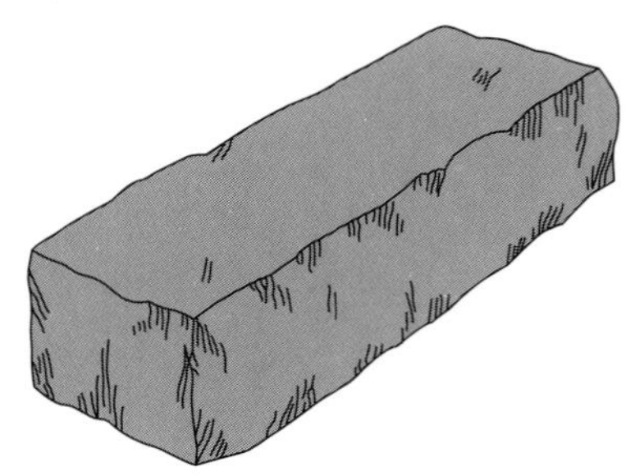

ASHLAR

A SLOPE GAUGE FOR TAPERED WALLS

To help you lay the stones of a dry wall for an inward taper, make a slope gauge from two 1-by-2s as long as the wall height—3 feet in this case. Fasten these arms together at one end with a single nail and pivot them so the distance between their outside edges at the free end equals 1 inch for each foot of wall height. Secure the free ends with a short two-piece wood block as shown at right. Finally, drive a second nail to lock the pivoting ends together.

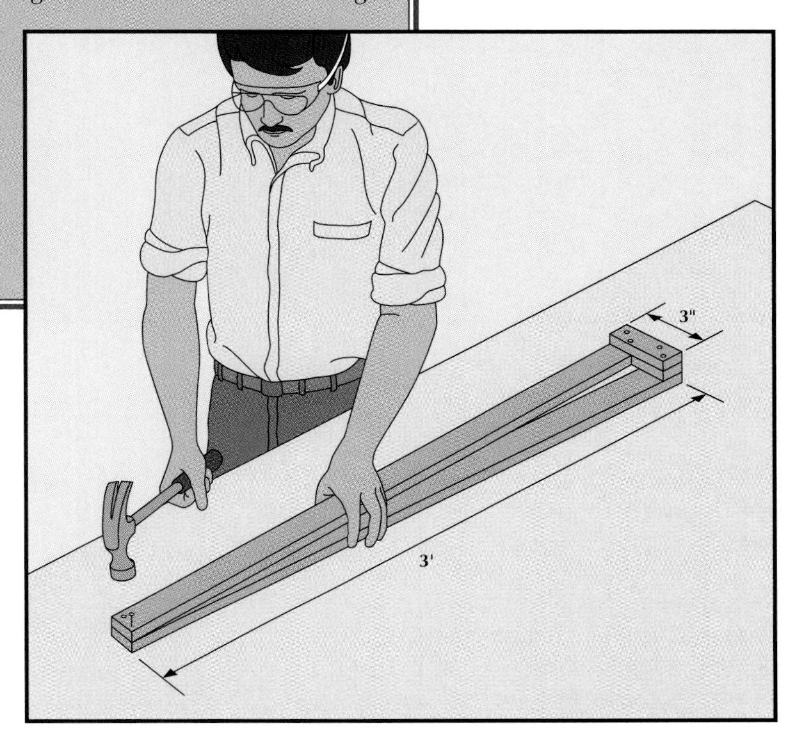

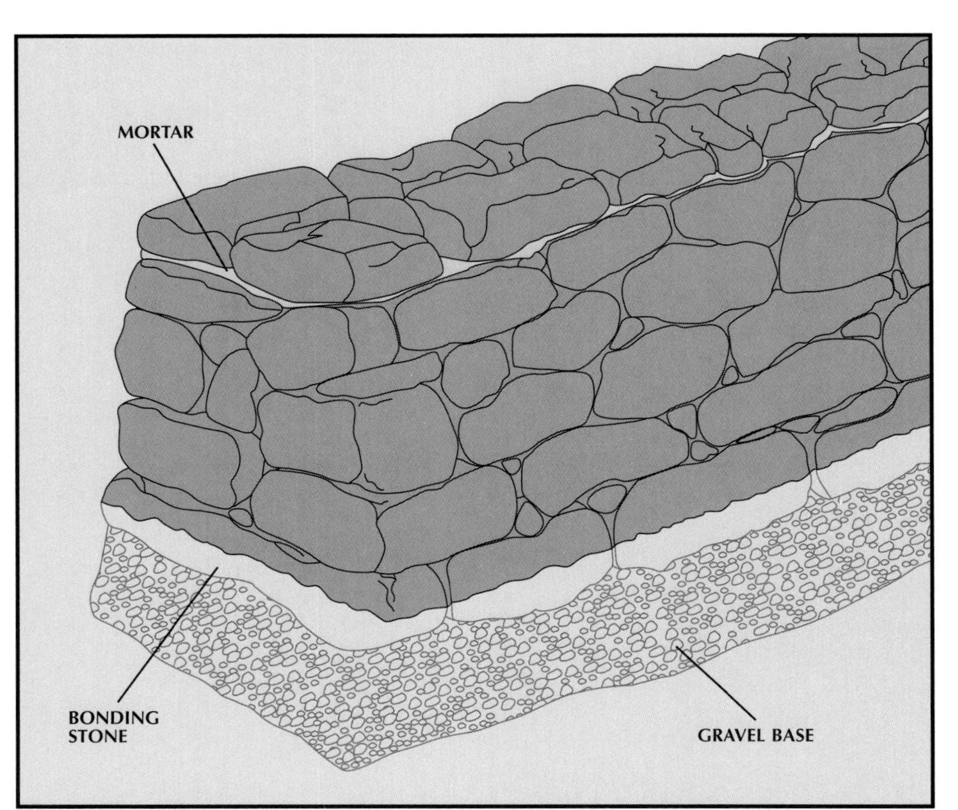

Anatomy of a dry wall. The wall sits in a 6-inch trench on a 5-inch layer of gravel. At one end of the base is a bonding stone that spans the width of the wall. Square-cut stones are laid flat, whereas rubble pieces tilt toward the center of the wall. The base is 2 to 3 feet wide, but each new course is slightly inset to taper the wall inward. Vertical joints in successive courses are staggered. The top layer of stones is often anchored in mortar to seal out water.

BUILDING WITH MORTARLESS STONE

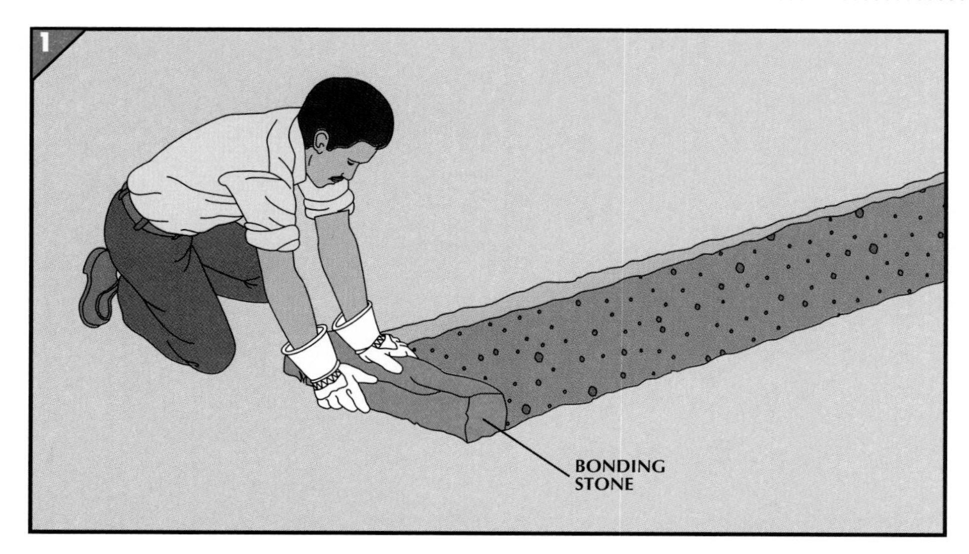

1. Setting the bonding stone.

◆ Dig a 6-inch trench to the length and width of the wall.
◆ Fill the trench with about 5 inches of gravel and tamp it down.
◆ Pick an even-faced stone as long as the wall width and set it at one end as a bonding stone *(left)*. If no stone is long enough to span the trench, lay two stones.

2. Laying the first course.

Use the largest stones for the first course; reserve the flattest ones for the top.
◆ Lay four or five stones along one side of the wall, setting them flat rather than on end or on their sides. Orient long pieces lengthwise along the edge of the trench, not across it *(right)*. Place the stones so any slope on the upper surface angles downward to the center of the wall. Alternate between large and small stones, thick and thin ones.
◆ Repeat along the other side of the trench.

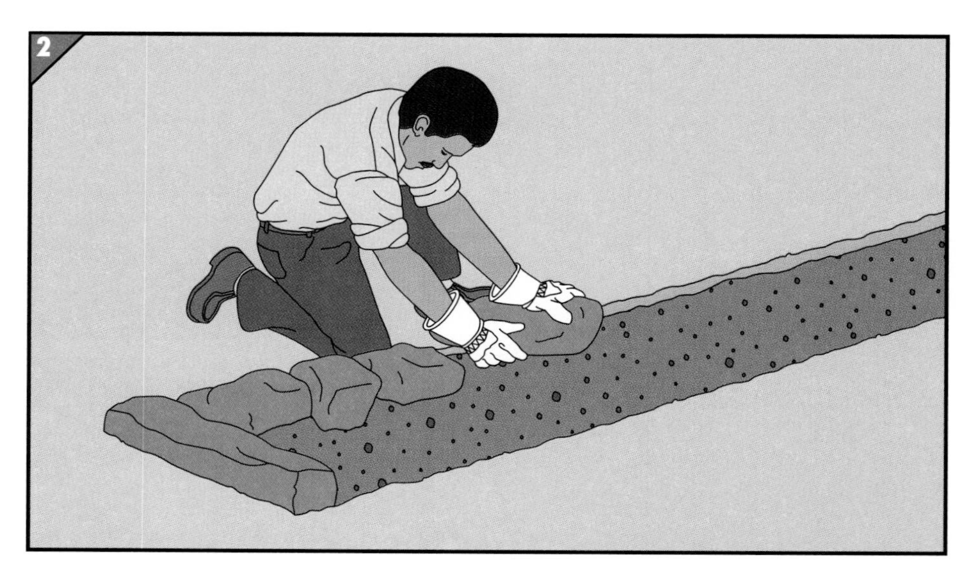

3. Filling in the middle.

◆ Fill in the center of the trench with small stones *(left)*, building up the middle section level with the edges.
◆ Continue laying stones along the sides and middle of the trench *(Steps 2 and 3)* until you reach the opposite end.

4. Laying the second course.

◆ Choose a stone long enough to overlap the bonding stone and the piece next to it in the first course.

◆ Place the stone along one edge of the wall so its top surface angles down slightly toward the center of the wall and its outside edge is set in a little from the underlying stone.

◆ Alternating from one side of the wall to the other, continue laying second-course stones that overlap joints in the first course, fit well with the stones underneath, and allow the top of the course to remain level.

◆ To check the inward taper of each stone, hold your slope gauge against the wall, and plumb it with a level *(left)*. If the stones do not align with the gauge, reposition them as necessary.

◆ Fill in the center of the course.

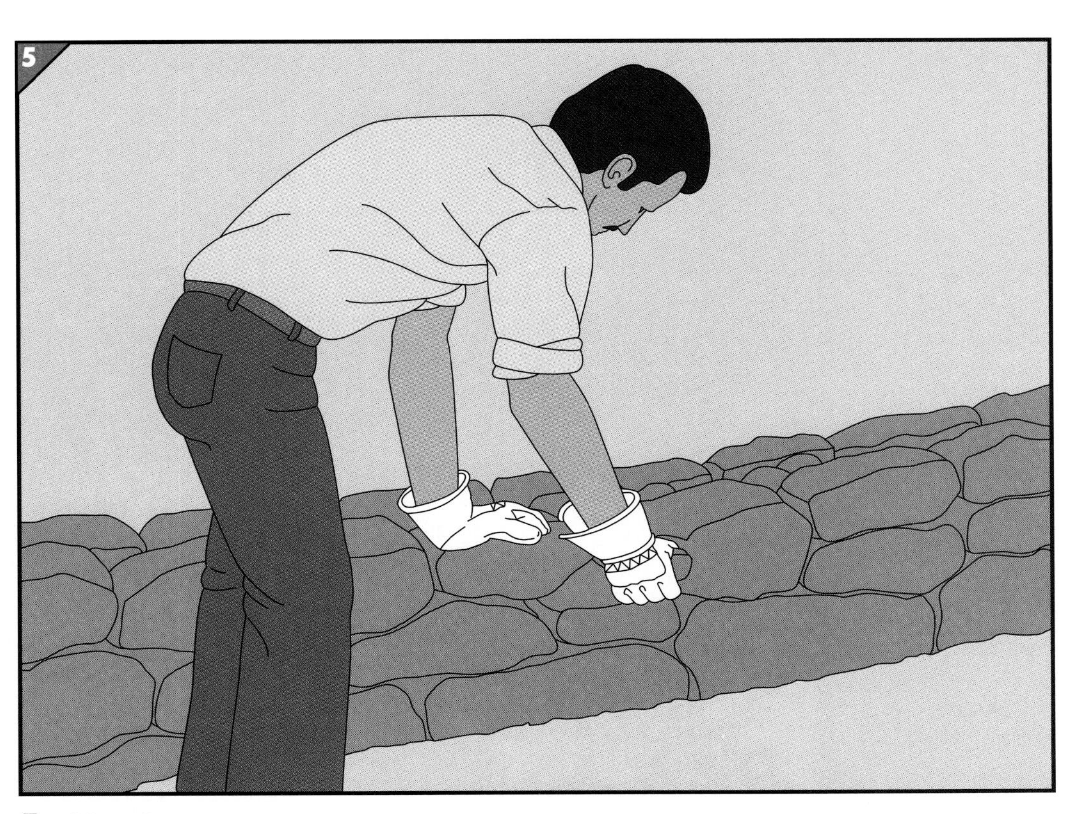

5. Shimming.

Lay the third course of stones in the same way as the second. Where a stone can be teetered from side to side, insert stone chips or small rocks as shims under an edge *(above)*, pushing them in so they are hidden. Ensure that the stone rests securely and tilts inward.

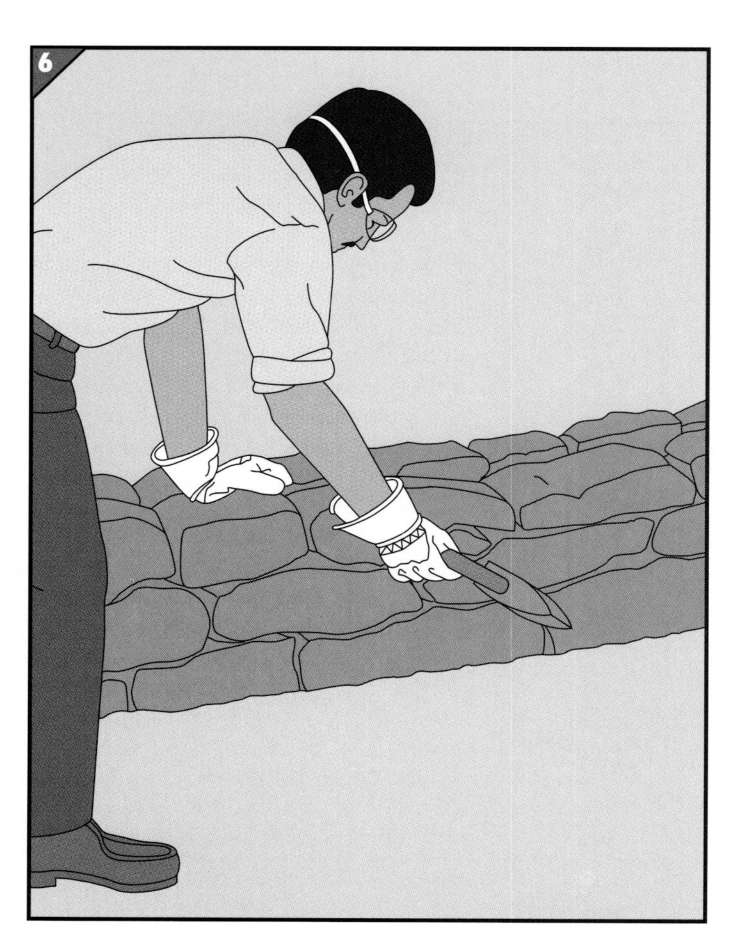

6. Chinking.

◆ Every three courses, fill in fist-sized or larger spaces in the wall by driving narrow stones into the gaps with a stonemason's hammer *(left)*. Leave smaller gaps alone.

◆ Keep each course level and check the taper of the wall with the slope gauge.

7. Mortaring the final course.

◆ Once you've laid the next-to-last course, prepare a batch of mortar *(pages 10-11)*.

◆ With a mason's trowel, cover the top course of stones with a 1-inch layer of mortar.

◆ Set the flat stones you reserved into the mortar *(right)*.

◆ Fill in the gaps between these stones with mortar, sloping the joints down from the middle out to the edges to prevent water from pooling on the wall.

◆ Trim excess mortar from the sides of the wall with the trowel.

◆ Once the mortar is thumbprint-hard, use a small piece of wood to rake the joints along the sides of the wall to a depth of about 1 inch.

DEALING WITH A CORNER

Interlocking the stones. Build up the first and second walls course by course as you would for a straight wall, but at the corner, use large stones (shaded area in walls at right) to make the 90-degree turn from the first wall to the second wall.

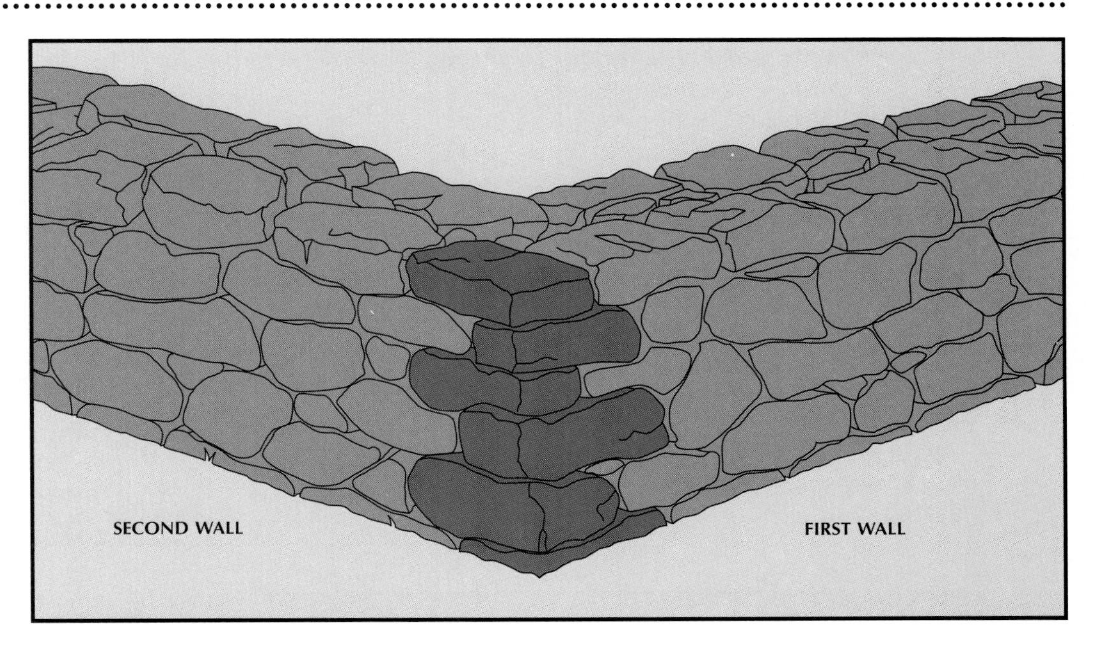

A MORTARED RETAINING WALL

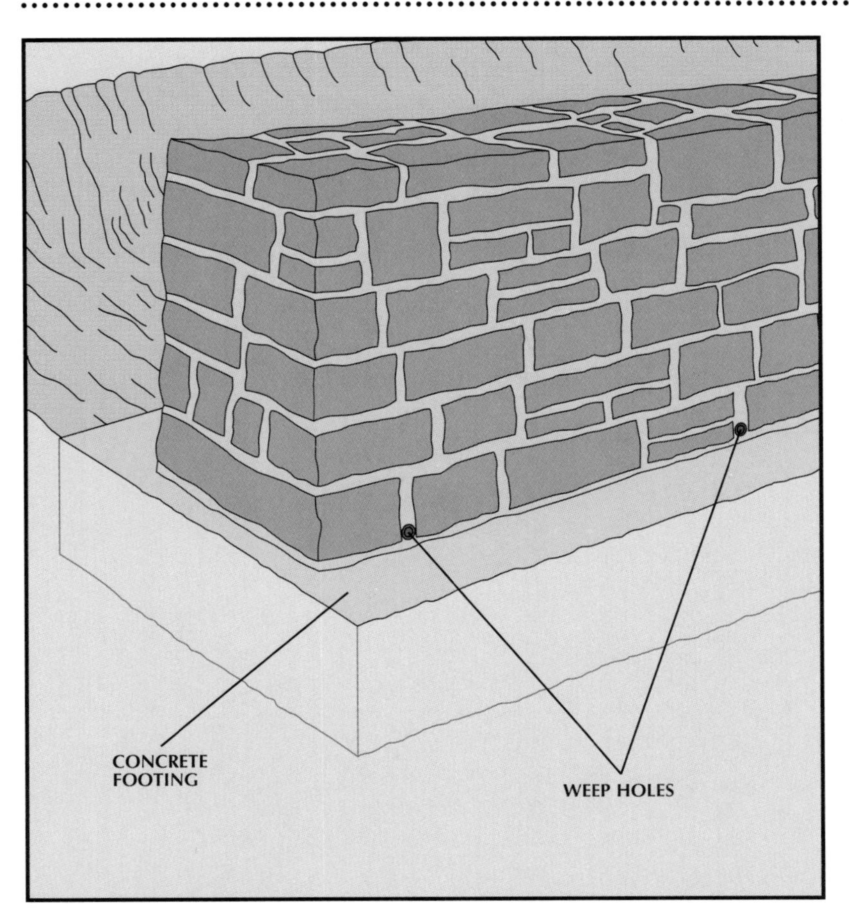

A stone retaining wall. The low, 18-inch-wide wall at left sits on a concrete footing 12 inches thick. The wall is built with square-cut stones laid out with $\frac{3}{4}$-inch-wide mortar joints and stacked so vertical joints are staggered as much as possible. Weep holes every 3 to 4 feet at the bottom lead water away from the soil behind the wall, and gravel banked in the trench behind the wall will enable water to seep down to the weep holes. If the soil is very fine, the fill can be kept clean by covering it with a layer of filter fabric, available at landscaping suppliers.

MORTARING THE STONE

1. Laying the mortar bed.

◆ Dig a trench for a concrete footing that will be about 5 inches larger on all sides than the wall. Behind the wall, dig the trench an additional 18 inches wide to accommodate the gravel fill.

◆ Cast the footing *(pages 63-64)* and let it cure for at least 48 hours.

◆ Mix mortar *(pages 10-11)* and spread a 1-inch-thick layer—in an area large enough for only a few stones—on the center of the slab *(right).*

◆ Pat the mortar but do not compact it—it should have a fluffy texture.

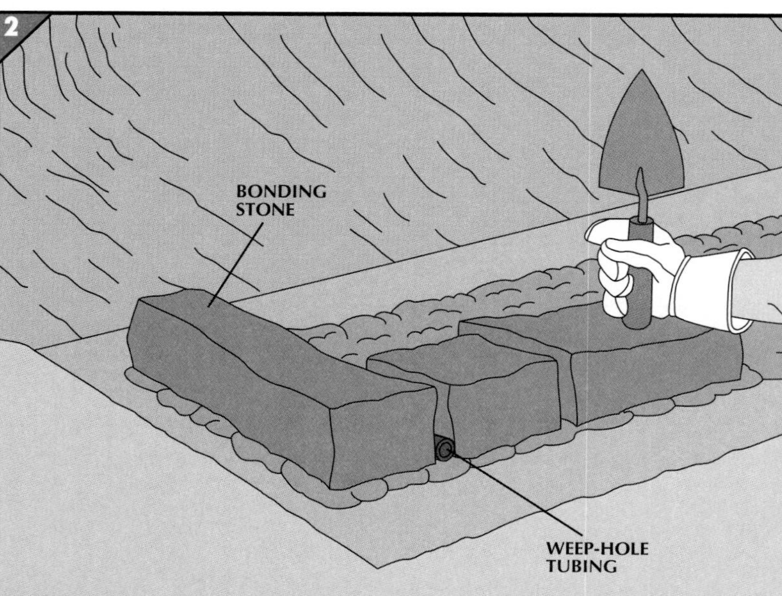

2. Setting the first course.

◆ At one end of the slab, set a bonding stone *(page 106, Step 1)* in the mortar.

◆ Place a length of 1-inch plastic tubing the width of the wall across the footing to create a weep hole.

◆ Lay the first course of stones along the front side of the footing, leaving a gap of about $\frac{3}{4}$ inch between units for mortar. Tap each stone down into the mortar with the trowel handle *(left).* Every 3 to 4 feet, lay weep-hole tubing across the footing into the gaps between stones.

◆ Set stones along the back of the footing, choosing ones that will fit on each side of the tubing.

◆ Finish laying the first course to the opposite end of the footing.

If you need to reposition a stone, wash off any adhered mortar before setting it in place again.

3. Filling in the middle.

◆ Fill in the center of the footing with tightly packed smaller stones, then cover them with mortar. Stack stones as necessary until the center is level with the courses on both sides *(right).*

◆ Once the mortar bed has stiffened, pack mortar into the outside joints between the stones around the perimeter of the footing.

◆ Lay successive courses the same way, checking the inward taper of the stones with a slope gauge *(page 107, Step 4).*

TRICKS OF THE TRADE

Props in a Rubble Wall

If you're building a mortared wall with rubble, the weight of large, irregular stones can squeeze mortar from joints. To support the stones, push small wood shims *(right)* under larger stones as you lay them. Once the mortar is hard, pull out the shims and fill the gaps with mortar.

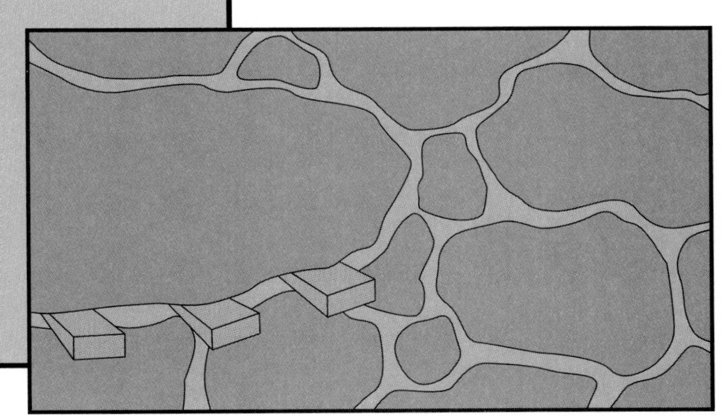

4. Raking the joints.

◆ On the top of the wall, lay and mortar a course of flat stones as for a dry wall *(page 108, Step 7)*.

◆ Rake the joints between stones with a small piece of scrap wood *(above)*, compacting the mortar and removing it to a depth of about 1 inch.

◆ After the mortar has set for a day or so, fill in behind the wall with gravel to within 6 inches of the top of the wall. Then cover the gravel with topsoil.

Concrete Block for Economical Construction

Concrete block gives the builder two major advantages over other types of masonry: economy and speed. A block wall costs less per running foot than a brick wall of an equivalent size, and takes about half as long to build. The gain over stone is many times greater.

Preparing to Build a Wall: Like a brick wall, a block wall is set on a concrete footing *(pages 63-64)*. Low walls—up to 3 feet high—are easiest to build. Taller walls are likely to require reinforcement *(page 117)*, although in some areas even low walls require reinforcement. Check your local codes before beginning the job.

The mortar for block walls is stiffer than that for bricks. Reduce the water in the mortar recipe *(pages 10-11)* so that the mixture will not simply slip from a trowel but must be shaken off.

Building with Blocks: Blocks come in a variety of shapes and sizes, each with a different function or design *(below)*. For optimal wall strength, lay them in the running-bond pattern *(page 72)*.

Blocks can be cut in the same way as bricks *(page 17)*; or you can use a circular saw fitted with a masonry blade or rent a masonry saw.

To seal the top of the wall from water, lay solid-top blocks for the last course, or fill the cores of the top-course blocks with mortar and add a coping of brick or $\frac{1}{2}$-inch flagstone *(page 114)*.

TOOLS

Mason's trowel
Mason's line and blocks
Mason's level
Tin snips
Convex jointer
Shovel

MATERIALS

Mortar ingredients: Portland cement, hydrated lime, masonry sand
Metal mesh
Thin-set mortar
Small flagstones

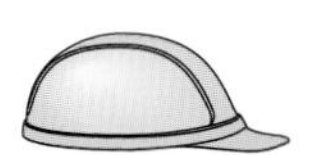

SAFETY TIPS

Mortar is caustic—wear gloves when working with it, and goggles and a dust mask when mixing it. Also don gloves when handling concrete blocks to protect your hands from the rough edges. Put on hard-toed shoes to prevent injury from dropped or falling blocks.

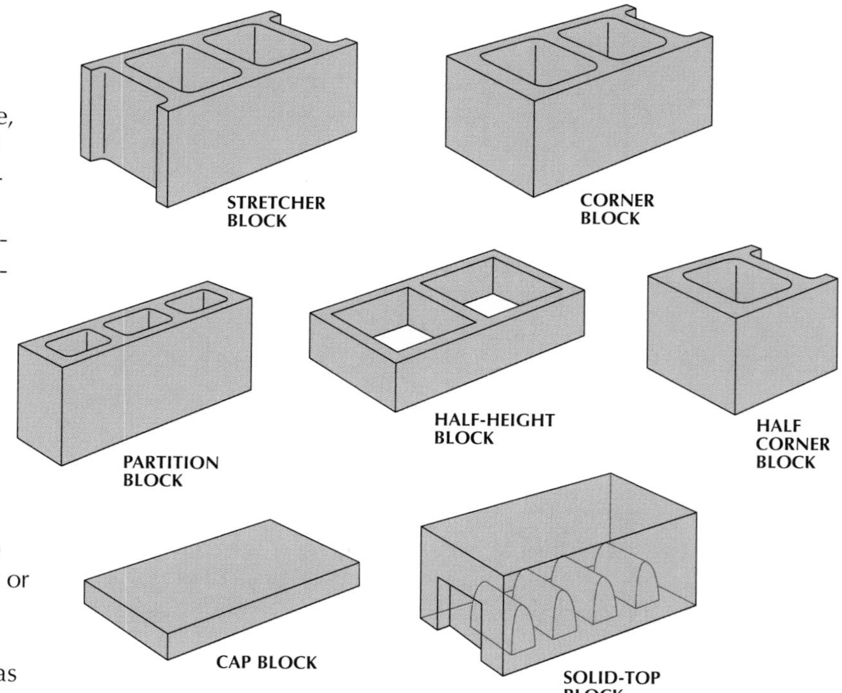

A block for every purpose.
Blocks of lighter-weight concrete are available, but a standard stretcher block weighs about 30 pounds and measures 8 by 8 by 16 inches. It has mortar-joint projections at both ends and two hollow cores bordered by partitions called webs. A variation of the stretcher, the corner block, is flat on one end for use at corners. Half-corner blocks laid at the end of every second course enable vertical joints to be staggered in a running-bond pattern; you can buy half blocks or cut them yourself from corner units. Partition and half-height blocks can be used as cores in brick structures *(page 121)*. Partition blocks can also provide weep holes in retaining walls *(page 116)*. Solid-top or cap units are handy for capping block walls. Special "architectural" blocks with different surface textures are also available, as well as screen blocks for decorative effect.

MORTARLESS BLOCKS

A variation on the standard concrete block is a type that does not need mortar. The unit at right, for example, has pins that slip into predrilled holes and lock the block to one above it. The block is tapered from front to back at each end, making it possible to build curved walls. In a straight wall, every other block is reversed.

ERECTING A BLOCK WALL

Laying the blocks.
If local codes dictate that the wall must be reinforced *(page 117)*, add steel reinforcements according to code specifications.

◆ For a wall that needs no reinforcement, first throw and furrow a bed of mortar $1\frac{1}{2}$ inches thick on the footing *(pages 11-12)*.
◆ Build stepped leads at the ends of the wall, adapting the technique for a brick wall *(pages 81-84)*, starting with a corner block at each end. For convenience, butter two blocks at a time: Stand them on end on the ground and spread mortar on their projections. Then, lay each block, lifting it by its outside webs and pushing it into the mortar bed and against the adjacent block, forming a $\frac{3}{8}$-inch mortar joint. Trowel away any extruded mortar. For subsequent courses, throw a mortar line two blocks long along the face shells of the previous course. Start and end every second course with a half corner block so vertical joints can be staggered. Periodically check for alignment and level as you lay the blocks.
◆ Once the leads are completed, fill in between them *(page 85, Steps 1 and 2)*, using a mason's line to keep the wall straight. For a wall longer than 20 feet, add control joints *(page 115)*.
◆ Rainproof the wall *(page 114)* one course from the top.
◆ When the mortar is thumbprint hard, finish the joints with a convex jointer *(page 14)*.

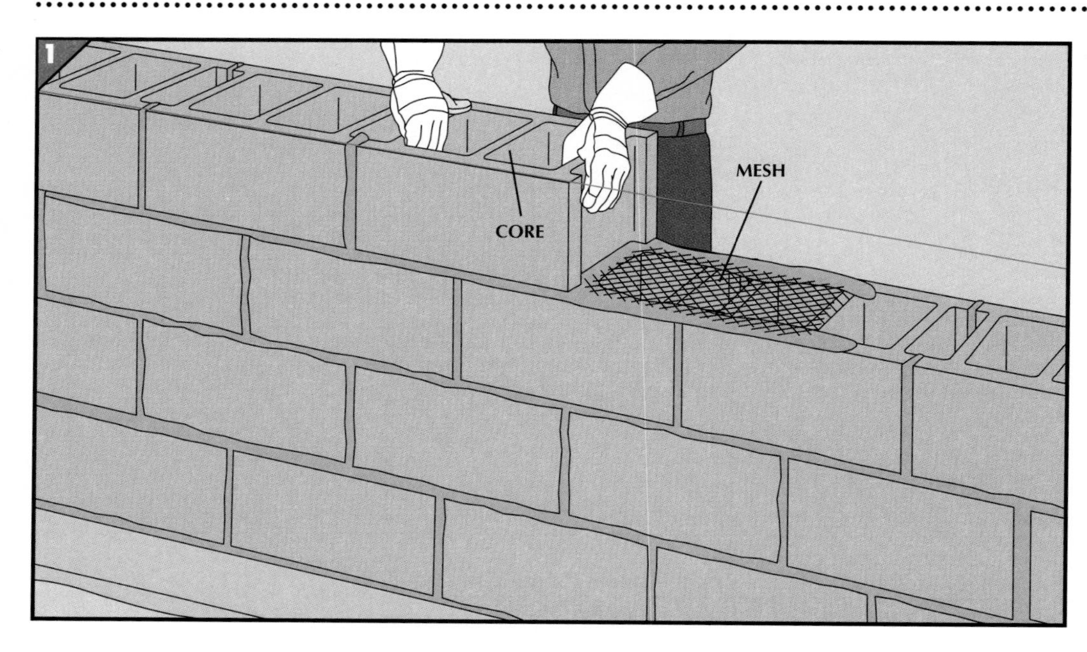

1. Filling the cores.

◆ Before throwing the mortar for the final course of blocks, use tin snips to cut metal mesh or hardware cloth into strips two blocks long and 1 inch wider than the cores.

◆ Apply the mortar for the top course and push the mesh into it.

◆ Lay the blocks on the mortar and mesh *(left)*.

◆ Trowel mortar into the cores, filling them even with the tops of the webs.

TRICKS OF THE TRADE

A Handy Funnel for Filling Blocks

To speed up the process of filling the last course of blocks with mortar, build a funnel from $\frac{1}{2}$-inch plywood. Cut the sides to the distance between the outside webs of a stretcher block and the ends Y-shaped to fit into the cores. To fill the blocks, use mortar thinned down just enough to be pourable.

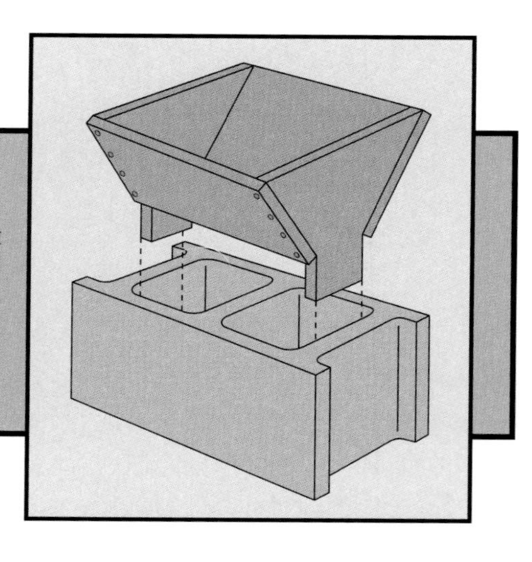

2. Adding a coping.

For increased weather protection, use thin-set mortar as for tile *(page 92)*.

◆ Trowel a $\frac{1}{2}$-inch mortar bed on top of the last course of blocks. In areas subject to freezing temperatures, spread the mortar higher along one edge of the wall than the other so the coping will slope and prevent water from pooling on the wall.

◆ Set small flagstones on the mortar, leaving $\frac{3}{8}$ inch of space between the pieces *(left)*.

◆ Fill the joints with mortar *(page 103, Step 5)*.

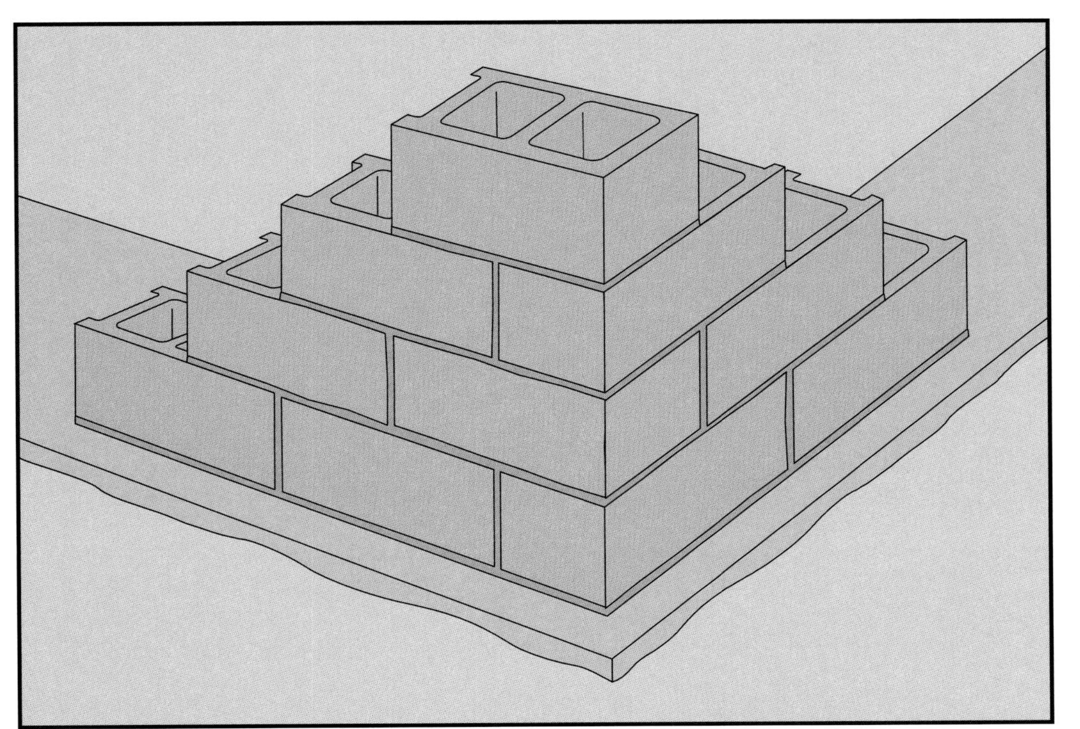

Corner leads.
For a wall with corners, first build corner leads as shown at left. The leads should step up half a block with each course, and the blocks at the corner should overlap. Once the leads are erected, fill in between them with blocks *(page 113)*.

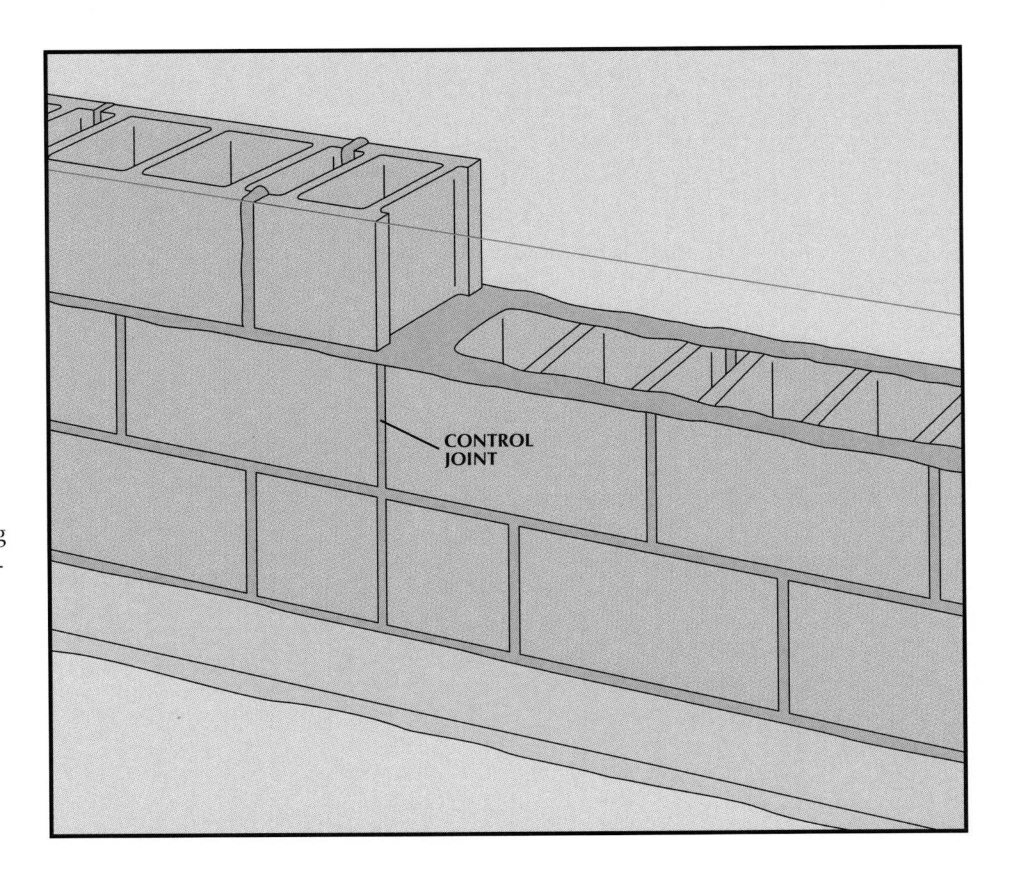

Control joints for long walls.
Block walls more than 20 feet long may develop cracks over time. Such cracking can be confined to weaker control joints located every 20 feet along the wall. To form the joints, use two half blocks in every other course, creating a continuous vertical joint *(right)*. Once the mortar stiffens, further weaken these joints by raking them to a depth of $\frac{3}{4}$ inch; fill the recess with caulk.

A BLOCK RETAINING WALL

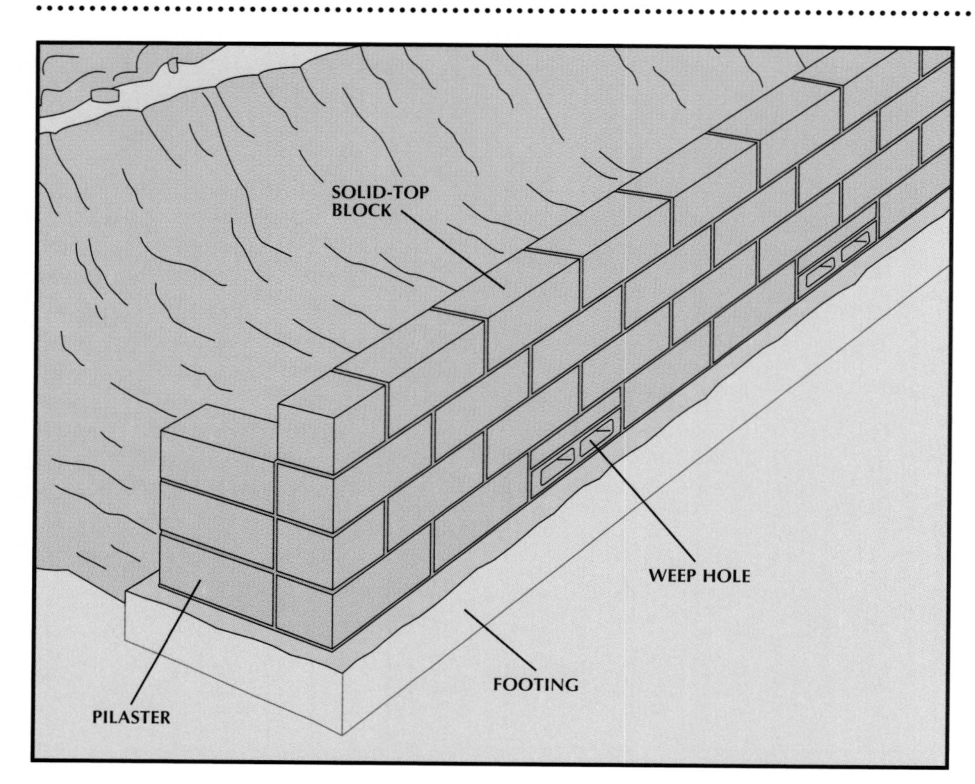

Anatomy of an earth dam.
To hold back the weight of soil, a block retaining wall must be buttressed by pilasters and provided with weep holes for drainage. In the wall at left, block pilasters are erected at each end of a concrete footing and at 10-foot intervals in between. One block shorter in height than the rest of the wall, the pilasters will be covered by the backfill. For drainage, gravel is banked against the back of the wall to allow water to seep down to the weep holes; then it is covered with 6 inches of topsoil. To keep the gravel clean in very fine soil, cover it with a layer of filter fabric (available at a landscaping supplier). Solid-top blocks complete the uppermost course of blocks, or standard blocks can be used, filled with mortar and decorated with a stone coping *(page 114)*.

RAISING THE STRUCTURE

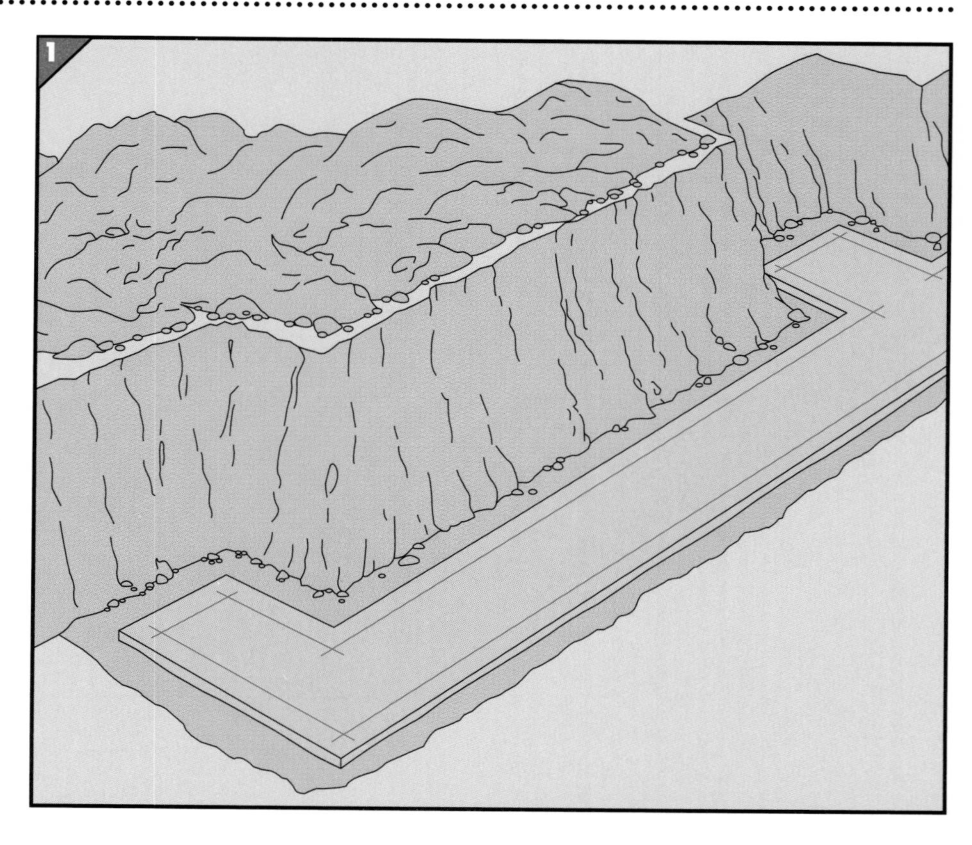

1. Preparing the site.
◆ Dig a trench about 2 feet wide, clearing $3\frac{1}{2}$-foot niches at the ends and every 10 feet in between for the pilasters. Pile the soil above the excavation so it will be easy to fill in later.
◆ Cast a footing for the wall *(pages 63-64)* with a 16-inch-square wing in the niches for each pilaster, as shown at right. Include reinforcing bars if necessary *(opposite)*.

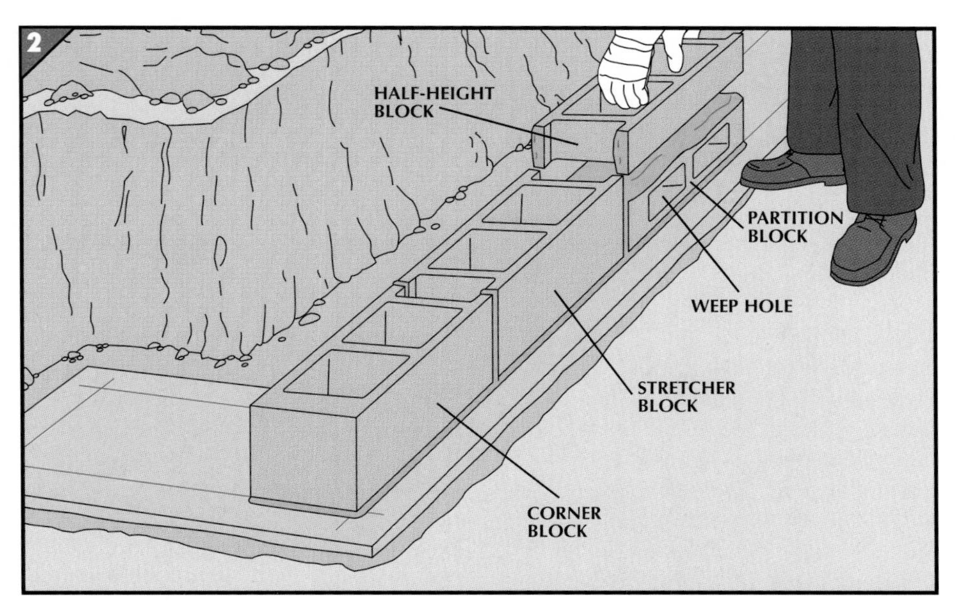

2. Starting the lead.

◆ Starting at one end of the footing, throw a mortar line for three blocks, then lay a corner block and a stretcher, followed by a partition block set on its side so the cores can serve as weep holes.

◆ To maintain the first course at a uniform height, spread mortar on the partition block, then butter one end of a half-height block and lay it on top *(left)*.

◆ Start another lead the same way at the other end of the footing.

3. Tying in pilasters.

◆ Lay a stretcher block at a right angle to the first block of each lead to form pilasters.

◆ Lay the second course of the leads and pilasters, then throw mortar beds for the third course of the leads and pilasters.

◆ Cut two pieces of metal mesh or hardware cloth 7 inches wide and about 16 inches long and push each one into the mortar on the pilaster and the first lead block, $\frac{1}{2}$ inch from the front of the wall *(right)*.

◆ Fill in between the leads, creating weep holes every third block in the first course and tying a pilaster into the wall at each projection in the footing. If weep holes coincide with a pilaster, offset the holes by one block.

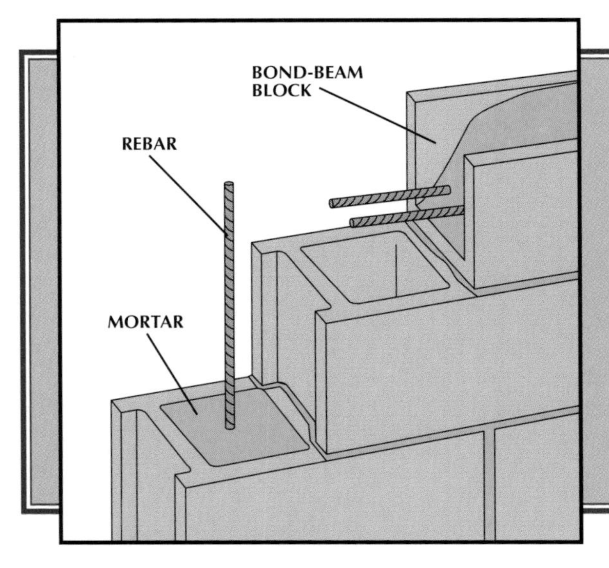

REINFORCING BLOCK WALLS

Depending on local codes, concrete-block walls more than a certain height need reinforcing. In areas with seismic activity, reinforcement is usually required for low walls as well. Requirements are generally even stricter for retaining walls. The size and placement of reinforcing bars (rebars) are also specified by code. For vertical buttressing, rebars are inserted down through the cores and tied with wire to rods cast into the footing. The cores are then filled with grout. For horizontal reinforcement, special webless bond-beam blocks are used. Rebars are set in these blocks, which are then filled with grout. Where rebars are spliced, the pieces should overlap by about 18 inches and be tied together with wire.

Stairs of Block and Brick

Steps made of concrete block are easier, quicker, and less costly to build than those made of poured concrete. Veneered with brick, tile, or stone, they are more attractive as well. And the techniques for assembling the bricks and blocks are the same as for other projects *(pages 82-87 and 112-114)*.

Plain vs. Veneered Steps: Block steps can be left exposed, or they can be painted or stuccoed. In these cases, the steps will be 8 inches high—the height of a concrete block; and the treads should be about 10 inches deep for a comfortable stride. If you want to veneer the steps with brick, tile, or stone, however, tailor the tread depth to the size of the veneer material to avoid having to trim the veneer to fit. With brick, for example, make the block treads 12 inches deep—one and one-half brick lengths plus the width of two mortar joints, as illustrated below.

Constructing the Slab: The blocks should rest on a 6-inch-thick concrete slab *(pages 54-58)* that is 4 inches wider and 2 inches longer than the steps and landing, including the veneer. Since the veneer will add to the height of the first step, the slab should be cast below grade by the height of one brick. To allow for drainage, slope the slab away from your house wall by $\frac{1}{4}$ inch for each foot of length. Place an expansion joint between the slab and the house foundation *(page 46)*.

Covering Existing Steps: When adding brick, tile, or stone veneer to existing block or cast-concrete steps, trim the veneer to fit and adjust mortar joints as necessary. To avoid a first step that is higher than the others, pave the walk leading to the steps.

TOOLS

- Chalk line
- Carpenter's square
- Mason's line and blocks
- Tamper
- Brick set
- Ball-peen hammer
- Mason's trowel
- Mason's level
- Convex jointer

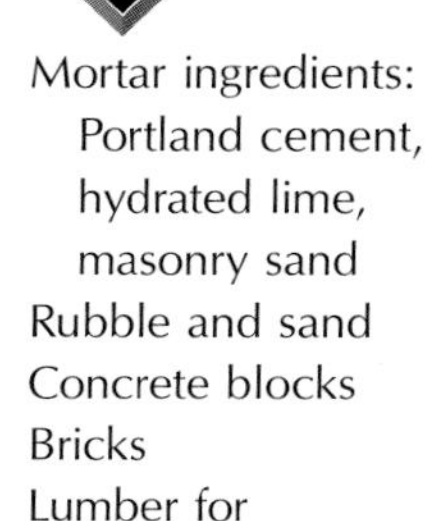

MATERIALS

- Mortar ingredients: Portland cement, hydrated lime, masonry sand
- Rubble and sand
- Concrete blocks
- Bricks
- Lumber for story pole

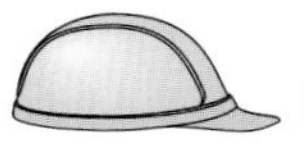

SAFETY TIPS

Mortar is caustic—wear gloves when working with it, and goggles and a dust mask when mixing it. Goggles are required when cutting bricks or blocks. Put on hard-toed shoes to prevent injury from dropped or falling blocks.

SHEATHING STEPS IN BRICK VENEER

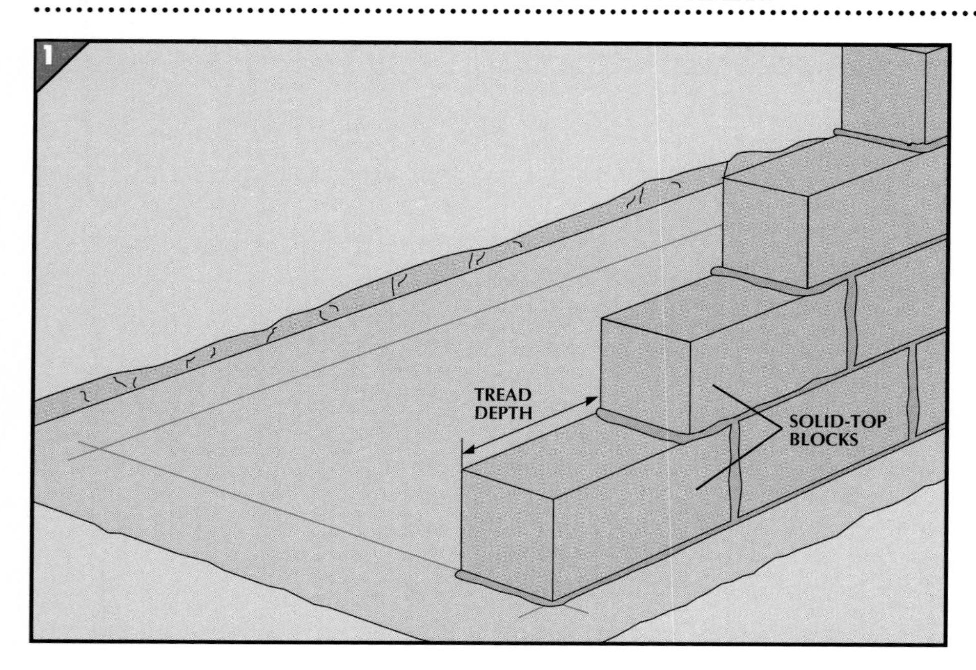

1. Laying the side walls.

◆ On the slab, outline the steps with a chalk line and carpenter's square.

◆ Starting at one front corner of the outline, lay a solid-top block. Extend the first course with stretcher blocks and end at the back with a corner block, trimmed to fit if necessary.

◆ Start the second course with a solid-top block. Offset it from the front of the first course by 12 inches, forming the first tread.

◆ Lay the remaining courses of the side wall this way *(left)*, then build the opposite side wall.

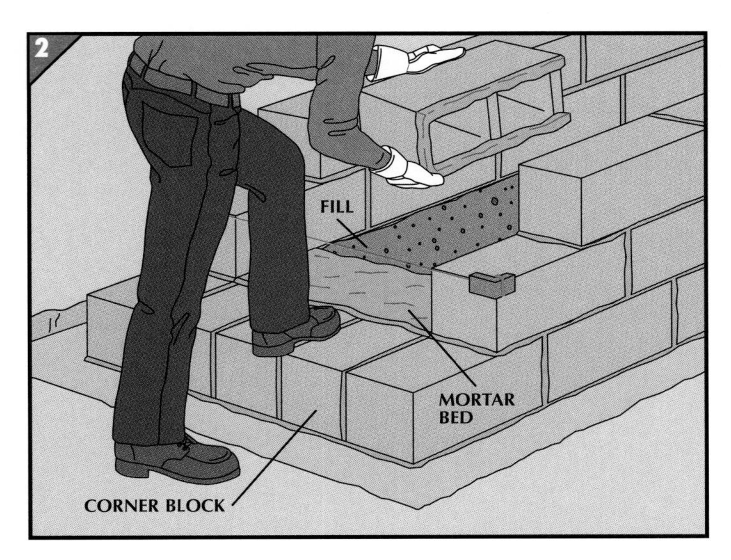

2. Filling in the treads.

◆ Between the front blocks of the side walls, stretch a mason's line and throw a mortar bed onto the slab.
◆ Butter one edge of a corner block, turn it on its side, and lay it against one side wall so its front end is flush with the mason's line. Lay additional corner blocks to complete the first tread.
◆ After the mortar has set for an hour, fill in the space behind the tread with rubble—scrap masonry—and sand. Tamp the fill even with the top of the first tread.
◆ To fill in the remaining treads, reposition the mason's line one step up, throw a mortar bed along the back portion of the preceding tread, and lay down corner blocks *(left)*. Fill in the landing the same way with stretchers laid on their sides.

3. Veneering the side walls.

◆ Start the first veneer course with a half brick *(page 17)*, positioned at one corner so the front of the next, full-length brick will be flush with the front of the first tread.
◆ Lay four courses of brick in a running bond pattern, mortaring them to the blocks. Check the height and alignment of the units with a story pole *(page 80)* and a level.
◆ Start each of the next three courses one and one-half bricks back from the first four *(right)*.
◆ Veneer both side walls up to the landing, offsetting each three-course set of bricks by one and one-half bricks from the one below.

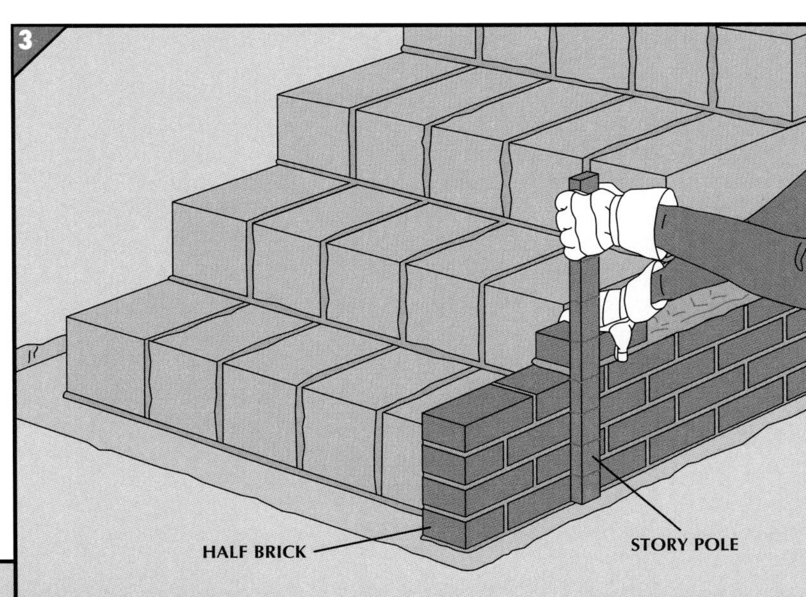

4. Paving the steps.

◆ Set up a mason's line even with the front end of the bottom course of brick veneer on the side walls.
◆ Lay bricks against the first riser flush with the mason's line, aligning vertical joints. As you lay the second and third courses, size the joints so the top course of bricks is level with the top of the first tread.
◆ Set bricks on the first tread so the front ends are flush with the outside edges of the riser bricks. Align the mortar joints and butter the ends of the second line of bricks that will contact the next riser up.
◆ Pave the remaining steps in this manner *(left)*, then cover the landing.
◆ When the mortar is thumbprint hard, finish the joints *(pages 14-15)*; use a $\frac{3}{4}$-inch convex jointer for the horizontal joints.

A Brick and Block Barbecue

Inexpensive concrete blocks can be covered with bricks to make a permanent freestanding barbecue. The construction job itself falls into two distinct stages: pouring a reinforced concrete slab *(pages 54-58)*, and laying the bricks and blocks *(pages 82-90 and 112-115)*.

Choosing the Site: First consult your local building inspector and fire department to see how codes or permit requirements will affect your plans. Locate the barbecue downwind from your home and neighboring houses, and from the dining area. Orient the structure so the prevailing winds will blow smoke away from the cook and create a draft for the fire.

Designing the Barbecue: The barbecue opposite rests on a concrete slab 8 inches thick and large enough to provide a 6-inch skirt around the back and sides and a 2-foot apron in front. You can tailor the design of the barbecue to suit your needs and tastes. Change the height of the grill, for example, by altering the number of brick courses and the height of the block core. You can adapt the shape to accommodate a larger, smaller, or second grill; or increase or decrease counter space: Lay out the first brick courses in the desired pattern and plan the core to fill the interior. It's also possible to add a door, converting the area below the ashpan into a warming oven or a storage area.

Accessories: The grill, fire grate, and ashpan are all available from masonry supply dealers, hardware stores, or restaurant equipment houses. The grill, made of cast iron or steel, should be rigid and heavy enough to resist sagging and being pushed out of position accidentally. The fire grate is usually made in small sections of heavy cast iron; you will probably need more than one section. Instead of separate grills and grates, you can buy a ready-to-install grate-and-grill system that enables you to adjust the cooking temperature by cranking the grill or grate up and down. The ashpan can be anything from a specially built sheet-metal container to a disposable aluminum broiler pan.

The stone countertop can be ordered cut to size from a mason supplier after the barbecue is constructed.

TOOLS

Tape measure
Chalk line
Carpenter's square
Mason's trowel
Mason's level
Mason's line
Hacksaw
Joint filler
Convex jointer
Straightedge
Utility knife
Paintbrush

MATERIALS

Bricks
Concrete blocks
Mortar ingredients (Portland cement, masonry sand, hydrated lime)
Lumber for story pole
Wall ties
Rebar ($\frac{3}{8}$")
Stone countertop
Resin-based stone sealing compound
Grill
Fire grate
Ashpan

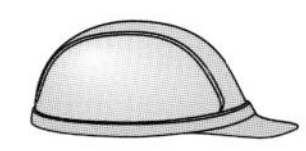

SAFETY TIPS

Mortar is caustic—wear gloves when working with it, and goggles and a dust mask when mixing it. Put on hard-toed shoes when handling bricks or concrete blocks.

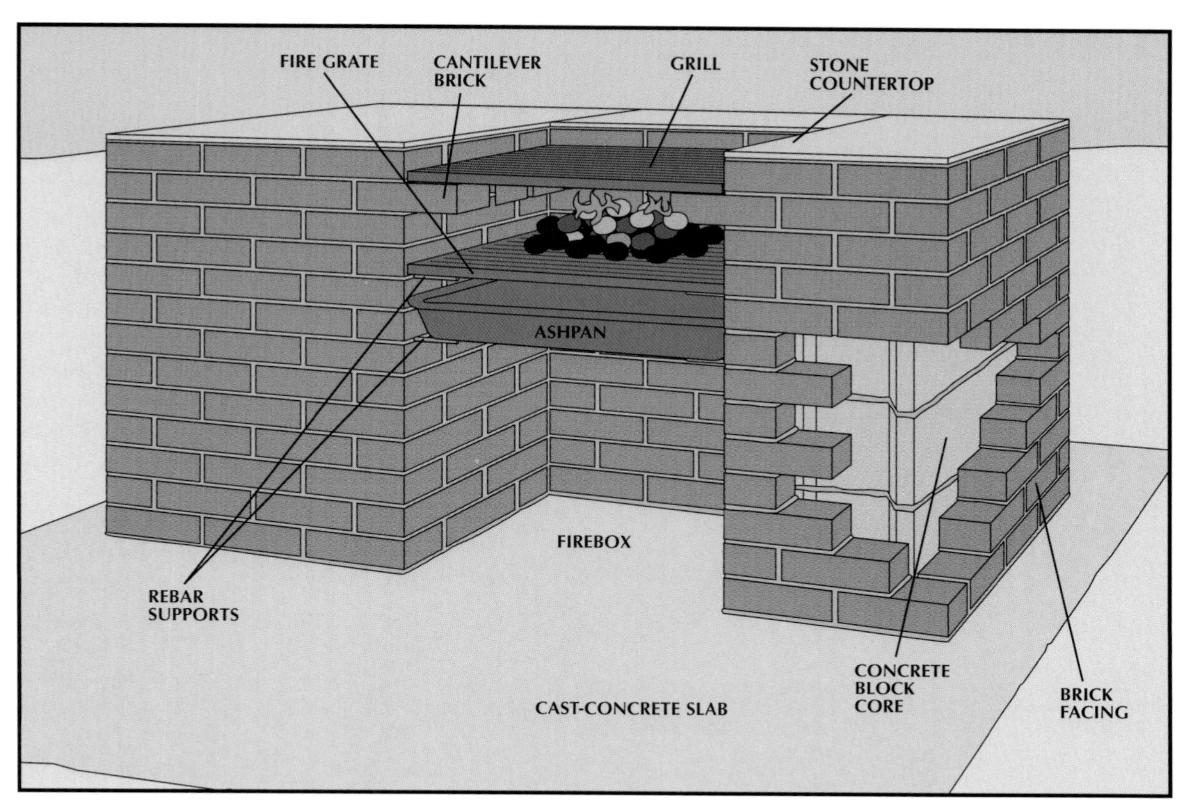

Antomy of a brick and block barbecue. The double-pedestal barbecue above measures about $5\frac{1}{2}$ feet wide by $2\frac{1}{2}$ feet deep and sits on a slab $6\frac{1}{2}$ feet wide and 5 feet deep.

Twelve brick courses in a running bond pattern with $\frac{1}{2}$-inch mortar joints will position the grill at about the same height as burners on a kitchen stove. The grill is held by cantilever bricks that project from each pedestal. The fire grate is positioned three brick courses below the grill, and the ashpan another two courses lower. Both the grate and ashpan are supported by stubs of rebar fixed in mortar joints.

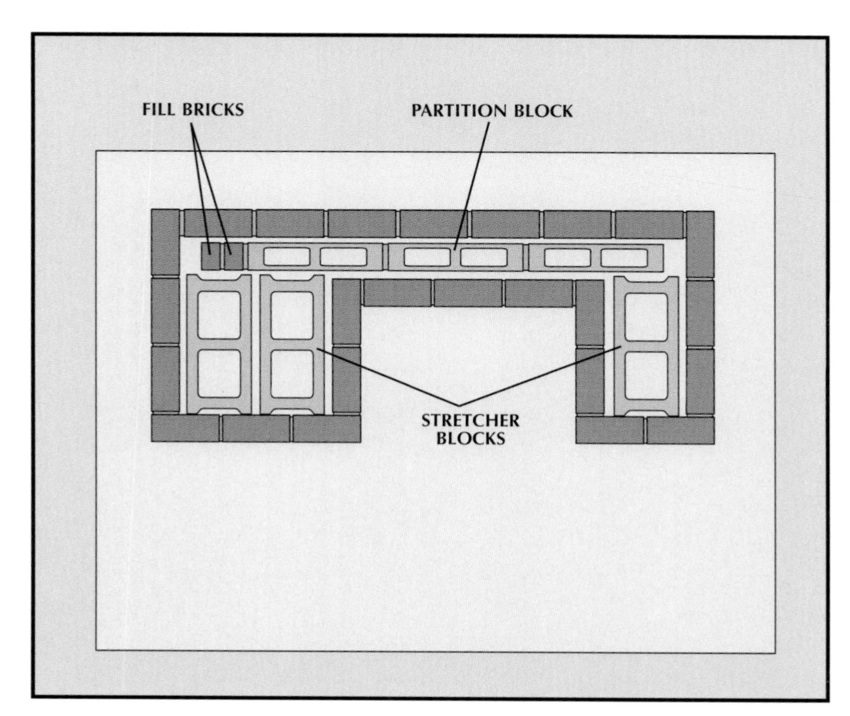

The pattern of bricks and blocks. The core of the barbecue consists of concrete blocks. In the two pedestals, stretcher blocks are laid directly atop each other, with vertical joints aligned, whereas the partition blocks at the back are laid in a running-bond pattern. Two bricks used for fill are mortared to one end of each course of partition blocks and are positioned at alternate ends to stagger the vertical joints. The block walls are not mortared to each other; the brick walls are anchored to the blocks with wall ties.

CONSTRUCTING THE BARBECUE

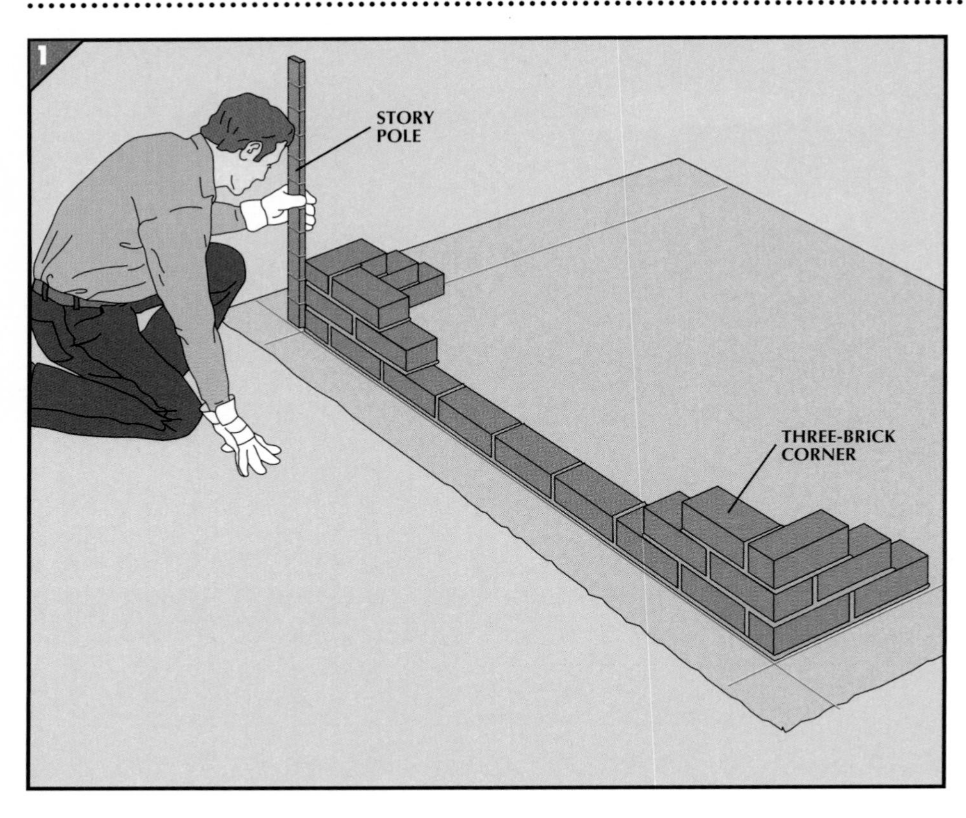

1. Starting the brick casing.

◆ Snap a chalk line for the back of the barbecue 6 inches from the edge of the slab. With a carpenter's square as a guide, chalk a perpendicular line 6 inches from one side of the slab.

◆ At the intersection of the lines, build a stepped three-brick corner *(pages 89-90, Steps 3-5)*. Check the brick height with a story pole *(page 80)*.

◆ Lay five more bricks to extend the first course along the back of the barbecue.

◆ Snap a third chalk line 6 inches from the other side of the slab, lay corner bricks along the line, then construct a second corner *(left)*.

◆ With a carpenter's square and a chalk line, outline the front of the barbecue and the firebox on the slab.

◆ Complete the first course of brick all around the barbecue and lay the second and third courses along the sides and back.

2. Starting the block core.

◆ Inside one of the pedestals, trowel just enough mortar onto the slab to form a bed for a stretcher block.

◆ Set a block on the mortar without displacing any bricks. Leave a $\frac{1}{2}$-inch gap between the block and bricks.

◆ Lay two more blocks in the pedestals.

◆ Starting $\frac{1}{2}$-inch from a back corner, set partition blocks along the back of the barbecue *(right)*.

◆ At the end of the course of partition blocks, set two bricks on end, mortaring them to the blocks as illustrated in the diagram on page 121.

◆ Check that the tops of the blocks are level with the top of the third brick course; adjust the thickness of the mortar beds under the blocks, if necessary.

3. Anchoring the bricks to the blocks.

◆ Lay the second and third courses of brick.

◆ Position wall ties every 10 inches across the blocks and bricks, placing them diagonally at the corners.

◆ Throw a mortar bed for a lead at one back corner of the barbecue. When you reach a tie, pick it up, apply the mortar *(above)*, and reposition the tie, pushing it into the mortar.

◆ Repeat Steps 1, 2, and 3 until you've laid three courses of blocks. When applying the mortar for the blocks, bend the ties out of the way without disturbing the brick, lay the mortar, then bend the ties back into the mortar.

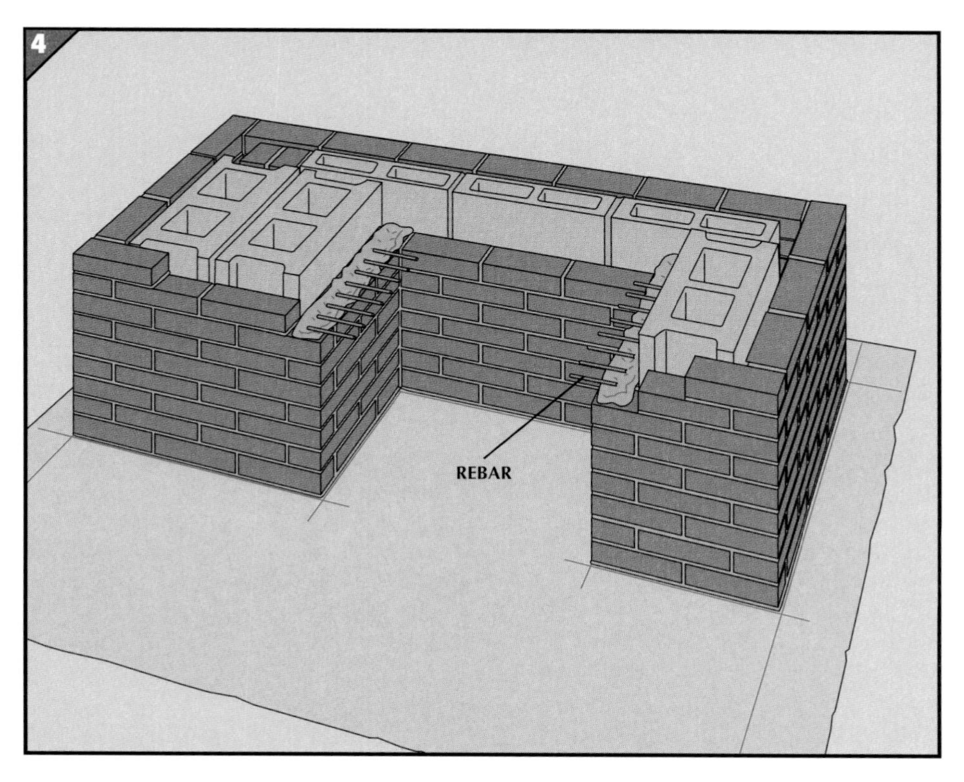

4. Embedding supports for the ashpan and grate.

◆ Before completing the eighth and ninth brick courses, cut $\frac{3}{8}$-inch rebar into pieces $7\frac{1}{2}$ inches long with a hacksaw.

◆ For the ashpan, apply the mortar for the eighth course of firebox brick; then, push eight cut rebars at equal intervals into the mortar on each side of the firebox so they project beyond the brick by 4 inches as shown at left.

◆ Finish laying the eighth and ninth courses of bricks.

◆ For the grate, embed rebars in the mortar between the ninth and tenth courses of firebox brick, then finish the tenth and eleventh courses.

5. Laying props for the grill.

◆ Begin laying the twelfth course of brick—when you reach the insides of the firebox, butter one edge of each brick, rather than an end, and set it into the mortar so the opposite end projects beyond the bricks below by 4 inches.

◆ To steady the projecting bricks while the mortar is setting, weigh them down with bricks placed on end *(right)*.

6. Trimming the joints.

◆ Once the mortar securing the projecting bricks has set for about 20 minutes, remove the weights and, with a $\frac{1}{2}$-inch joint filler, rake excess mortar from the joints between them *(above)*.

◆ Position the last set of wall ties, then lay the thirteenth course of brick around the top of the barbecue.

7. Paving over the blocks.

◆ Lay a mortar bed on the fourth course of blocks.
◆ Fill in with a layer of brick laid in the pattern shown at right, buttering both ends and one side of each brick.
◆ Once the mortar is thumbprint hard, finish the joints *(pages 14-15)*.

8. Topping the barbecue.

◆ On one side of the barbecue top, place a large piece of cardboard, weigh it down, and outline as much of the top as possible on the underside of the cardboard *(left)*.
◆ Remove the cardboard, complete the outline with a straightedge, then cut along the lines to make a template. Reposition the template on the top and trim it flush with the edges.
◆ Make templates for the other side and the center section of the top.
◆ Trim $\frac{1}{2}$ inch from each end of the center template to allow for mortar joints in the stone top.
◆ At a masonry-supply yard, have a piece of stone cut to match each template. Mortar them to the top of the barbecue.
◆ Let the mortar dry for two days, then coat the stone and its mortar joints with a resin-base stone-sealing compound to protect it from grease spatters.

INDEX

Time-Life Books is a division of Time Life Inc.

TIME LIFE INC.
PRESIDENT and CEO: George Artandi

TIME-LIFE BOOKS
PRESIDENT: John D. Hall
PUBLISHER/MANAGING EDITOR: Neil Kagan

HOME REPAIR AND IMPROVEMENT: Masonry
EDITOR: Lee Hassig
MARKETING DIRECTOR: James Gillespie
Deputy Editor: Esther R. Ferington
Art Director: Kathleen Mallow
Associate Editor/Research and Writing: Karen Sweet
Marketing Manager: Wells Spence

Vice President, Director of Finance: Christopher Hearing
Vice President, Book Production: Marjann Caldwell
Director of Operations: Eileen Bradley
Director of Photography and Research: John Conrad Weiser
Director of Editorial Administration: Judith W. Shanks
Production Manager: Marlene Zack
Quality Assurance Manager: James King
Library: Louise D. Forstall

ST. REMY MULTIMEDIA INC.
President and Chief Executive Officer: Fernand Lecoq
President and Chief Operating Officer: Pierre Léveillé
Vice President, Finance: Natalie Watanabe
Managing Editor: Carolyn Jackson
Managing Art Director: Diane Denoncourt
Production Manager: Michelle Turbide

Staff for *Masonry*

Series Editors: Marc Cassini, Heather Mills
Series Art Director: Francine Lemieux
Art Director: Robert Paquet
Assistant Editor: Rebecca Smollett
Designers: François Daxhelet, Jean-Guy Doiron, Robert Labelle
Editorial Assistant: James Piecowye
Coordinator: Dominique Gagné
Copy Editor: Judy Yelon
Indexer: Linda Cardella Cournoyer
Systems Coordinator: Éric Beaulieu
Other Staff: Lorraine Doré, Geneviève Monette

PICTURE CREDITS
Cover: Photograph, Robert Chartier. Art, Robert Paquet.

Illustrators: Gilles Beauchemin, Michel Blais, Adolph E. Brotman, Nick Fasciano, Peter McGinn, Roger Metcalf, Jacques Perrault, Ray Skibinski, Ed Vebell, Whitman Studio.

Photographers: **End papers:** Glenn Moores and Chantal Lamarre. **8, 9, 14, 21, 43, 58, 71, 74, 83, 92, 94, 100, 113:** Glenn Moores and Chantal Lamarre. **27:** Stone Construction Equipment. **29:** Brick Institute of America (efflorescence, brown stain, rust); Robert Chartier (smoke, oil, paint). **30:** DeWalt Industrial Tool Company Inc. **34, 35, 52:** Portland Cement Association. **56:** Wacker Corporation.

ACKNOWLEDGMENTS
The editors wish to thank the following individuals and institutions: B&S Centre de Toiture Ltée., Montreal, Quebec; Bon Tool Co., Gibsonia, PA; Brick Institute of America, Reston, VA; Briqueterie St-Laurent, La Prairie, Quebec; Canada Brick, Streetsville, Ontario; Cooper Tools, Barrie, Ontario; Crossville Ceramics, Bedford, TX; DESA International, Bowling Green, KY; DeWalt Industrial Tool Company, Mississauga, Ontario; DG Centre de Renovation, Montreal, Quebec; Louis V. Genuario, Genuario Construction Co., Inc.; Laticrete International, Aurora, Ontario; Marshalltown Trowel Company, Marshalltown, IA; Mason Contractors Association of America, Oak Brook, IL; Matco Ravary Inc., St-Leonard, Quebec; Jess McIlvain, Jess McIlvain and Associates, Bethesda, MD; National Concrete Masonry Association, Herndon, VA; National Tile Contractors Association Inc., Jackson, MS; Portland Cement Association, Skokie, IL; Simpson Strong-Tie, Pleasanton, CA; Stone Construction Equipment, Inc., Honedye, NY; Versa-Lok Retaining Wall Systems, Oakdale, MN; Wacker Corporation, Menomonee Falls, WI; Fred Waller, Raleigh, NC

First printing. Printed in U.S.A.
Published simultaneously in Canada.
School and library distribution by Time-Life Education, P.O. Box 85026, Richmond, Virginia 23285-5026.

Library of Congress Cataloging-in-Publication Data
Masonry / by the editors of Time-Life Books.
p. cm. — (Home repair and improvement)
Includes index.
ISBN 0-7835-3906-1
1. Masonry—Amateurs' manuals.
I. Time-Life Books. II. Series.
TH5313.M35 1996
693'.1—dc20 96-43206